JOURNAL OF
INDUSTRIAL ENGINEERING
AND MANAGEMENT SCIENCE

Volume 1, 2016

JOURNAL OF INDUSTRIAL ENGINEERING AND MANAGEMENT SCIENCE

Editor-in-Chief: Kuinam J. Kim, Kyonggi University, South Korea

This journal is published in collaboration with the iCatse organization: www.icatse.org

Aims and Scope
Journal of Industrial Engineering and Management Science is a peer-reviewed, Open Access journal that publishes theoretical and empirical research articles in all areas of industrial engineering and management Science. The journal covers the following subject areas, but is not limited to:

- Decision Analysis and Methods
- Information Processing and Engineering
- Intelligent Systems & Manufacturing Systems
- Healthcare Systems and Management
- Quality Control and Management
- Reliability and Maintenance Engineering
- Finance and Risk Management
- Operations Research
- Production Planning and Control
- Engineering Economy and Cost Analysis
- Supply Chain Management
- Marketing Science

Published, sold and distributed by:
River Publishers
Alsbjergvej 10
9260 Gistrup
Denmark

River Publishers
Lange Geer 44
2611 PW Delft
The Netherlands

Tel.: +45369953197
www.riverpublishers.com

ISSN 2446-1822 (Online Version)
ISBN 978-87-93519-54-1

JOURNAL OF INDUSTRIAL ENGINEERING
AND MANAGEMENT SCIENCE

Volume 1, 2016

Knowledge Sharing and the Innovation Capability of Chinese Firms: The Role of Guanxi

Oswaldo Jose Jimenez Torres

School of Business Management, Harbin Institute of Technology, Harbin, China
E-mail: Oswaldojim@gmail.com

Received 20 July 2016; Accepted 10 September 2016;
Publication 30 September 2016

Abstract

Building on the theory of social capital the present theoretical study aims to introduce a framework to analyze the extent to which the innovation capability of Chinese firms is affected by Guanxi as a moderator between the former and the knowledge sharing behaviors among individuals and teams. The three dimensions of knowledge sharing considered here are: Type of knowledge, quantity and quality of knowledge shared among employees. Also this article includes propositions and recommendations to lead future research in this area.

1 Introduction

The set of social principles known as Guanxi (关系) is the central piece of the Chinese society, it has its origins in the Confucianism which has been part of the Chinese social system for over 2.500 years. Confucianism encloses the theory of relationalism, which states that individuals must be relationship-based. Even though, this ancient philosophy also clearly differentiates the most important human relationships known as *lun* or *wu lun* (五伦) which can be translated as (Five Circles or Five Cardinal Relationships), and those are: ruler-subject (君臣), father-son (父子), husband-wife (夫婦), elder

Journal of Industrial Engineering and Management Science, Vol. 1, 1–18.
doi: 10.13052/jiems2446-1822.2016.001

brother-younger brother (兄弟), and friend-friend (朋友). In its core Confucianism have three main precepts, which are: The importance of human relationships, the social order and the pertinent moral principles that guide the behaviors of individuals [1]. Much of the literature on Guanxi is business-orientated, in that, is focused on grasp and explain to non-Chinese the relevance and complexities of this concept for conducting negotiations and achieve competitive advantage in China as well as dealing with Chinese organizations overseas. While the values of the Chinese society have evolve trough out the years it's still based on the idea of the Confucian society and so Guanxi relationships are conducted considering those values. Two core principles of Guanxi are identified in literature. The first one is the principle reciprocity or exchange of favors (*renqing*), this means that individuals must assist those who assisted them, and it's important to notice that this reciprocity must be balanced. The second, it's the principle of long-term equity, it means that parties of an exchange are permitted to share the outcome of the exchange giving their input on it in the long-term [2].

Different authors have studied the effects of social norms in organizations, such as [2], whom based on the theory of social capital proposed that favor exchanges among colleagues in organizations outside the private lives frequently implicate the use of organizational resources as well as positions. This practices benefit the subjects involved in this dynamics, in that, the provider increases his private social capital and the receiver coup his needs. Notwithstanding, those behaviors doesn't contribute to enhance the public social capital of the organization as a whole. [3] showed that individuals who share strong empathy with each other tent to foster this empathy by providing resources also between them, even when this implicates neglecting the collective good [4]. In the light of this, it can presumed that the behaviors implicit in Guanxi relationships, even though generate benefits for individuals whom share close ties with each other, might be detrimental for organizations, specifically for knowledge sharing practices among team members, because those would act as a barrier restricting the equalitarian or unbiased flow of knowledge between all the members involve in specific tasks for an organization and therefore holding back the innovation capability of the firm. The role of intra-organizational knowledge sharing in enhancing innovation capability of firms has been the focus of previous research for instance [5, 6], its importance in accelerating individuals and teams problem-solving capacity, generation of new ideas and business models is created by both knowledge donating and collecting among individuals in an organization.

The clear distinctions established on the Confucian precepts about the kind of relationships individuals have, the obligations of the parties and the moral principles applicable in each different case suppose and unequal or discriminatory treatment of members in society or groups. Considering this, it's worthy to analyze the impact of all this socio-cultural connotations in the organizational context in order to identify whether those are beneficial, harmful or irrelevant for organizations. Therefore the main assumption of the present study is that the innovation capability of Chinese organizations is affected by the role of Guanxi as a moderator between the former and the knowledge sharing behaviors of their staff. Hence this study presents a conceptual framework built on previous literature as well as propositions and directions for upcoming research on this subject. The dimensions of knowledge sharing to be addressed are the following: Type of knowledge (explicit and tacit), quality and quantity.

2 Literature Review

2.1 Guanxi and the Theory of Social Capital

The basic notion embedded in the theory of social capital is that a person's social networks, conformed by his (friends, colleagues and other contacts) constitutes a form of capital and that individuals often times engage in activities leading to enhance their private social capital regardless of the public social capital. Therefore after observing the principles of Guanxi relationships and the behaviors of individuals whom follow those precepts distinguishing between insiders and outsiders to their social circles, academics have identified the theory of social capital as a fit kaleidoscope to analyze this social norm with due scientific rigorosity. [7] studied the concept of social capital considering the approach of two main intellectual streams focused on social action. One of those treats actors as social individuals highly influenced by a set of norms, rules as well as obligations. On the other hand, we found the economical stream of though, the basic assumption here is that the actor drives himself according only to his self-interest. [8] explained that social capital can operate in two forms depending on the actors and their dynamics, in this sense, public social capital is owned by the unit or organization as a whole while private social capital is owned only privately by a certain group of individuals. [9] defined social capital as: "The sum of the resources, actual or virtual, that accrue to an individual or a group by virtue of possessing a durable network of more or less institutionalized relationships of mutual acquaintance and recognition".

[10] introduced a comprehensive analysis on the different definitions that had been attributed to Guanxi throughout decades, also some examples of the various meanings the word Guanxi can have depending on the context. Some definitions relevant for academia are: "Friendship with implications of a continual exchange of favors" [11]. "The concept of drawing on connections in order to secure favors in personal relatives" [12]. Other approaches taken for the study of Guanxi are [13], they focused on explain the differences between the communist based managerial system applied in China and the new one managers are engaging on nowadays which is a capitalistic social system with clear Anglo-Saxon features, for instance strong ethical values and clear rules of trade, [2] this is a conceptual article in which the authors based on the theory of social dilemma explained how the close ties implicit in Guanxi relationships generate "externalities" in groups and also explained some managerial actions that mitigate those, such as foster group identity in organizations. [14] the focus of this research was to examine indigenous forms of informal influence in different locations. The authors analyzed the concepts and dynamics of Guanxi, Wasta, Jeithinho and Svyazi, they verified that the dynamics implicit in this concepts are more frequent in places with self-enhancement values, low self-transcendence values and high impunity of corruption. [15] this work provides a much needed comparison between Guanxi and its analog in the Arab world, Wasta. Here the author explained that one of the main similarities that makes the concepts of Guanxi and Wasta to flourish in their respective societies is that those are low-context societies. [16], in that article authors studied which factors could foster knowledge sharing, the selected factors were: Trust, Guanxi orientation and face. [17], here the concept of Guanxi was operationalized, presenting the following components: Trust, relationship commitment, and communication. [18], the dynamics and importance of Guanxi are studied not also at the corporate level but also to the governmental level.

[12] focused on analyze how firms apply Guanxi, specifically the way it is used to cope with competitive and resource disadvantages. Based on their results, they explained that Guanxi relationships might benefit organizations in terms of market expansion and competitive advantage but not in internal operations. [19], this work studied the ethical dimension of Guanxi, here some of the negative features of it were presented, such as corruption and misgiving behavior. In his final remarks the author argued that in the case of international firms Guanxi it's still necessary, specially at the beginning of their operations in China, but its importance is likely to decline as companies

start acting according to pure market strategy, finally the author anticipated the total misused of Guanxi in the feature due to China's modernization.

2.2 Knowledge Sharing

Previously, authors have studied knowledge sharing with the following approaches: [20] concentrated on explore the extent to which cultural factor affect knowledge sharing in virtual communities. Based on their findings, they argued that the issue of saving face in China was less relevant than expected. Hence, considering arguments of [21, 22] on the evolution of Guanxi, it could be presumed that this discrepancies between the verified and expected importance of saving face as an element associated with Guanxi practices might was caused by an evolution of the Guanxi itself. Also, [23] identified major barriers to knowledge sharing in China, such as modesty requirements and competitiveness. On the other hand those issues were less important in Russia and Brazil. [24] explained how personal networking and membership affects knowledge sharing also in Russia and China.

Authors also have identified two dimensions of knowledge sharing, even though each study has defined those dimensions using different concepts, in essence, those can be reduced to giving and receiving knowledge. [25] explained that the knowledge sharing process involves bringing or ("donating") knowledge and getting or ("collecting") knowledge. Moreover, [26] while focusing on the knowledge sharing dynamics in virtual communities proposed that both the demand and supply of knowledge are critical for such communities to be vibrant. Therefore, aiming to assess the knowledge sharing among employees in the presence of Guanxi, the present study proposes to analyze both dimensions of knowledge sharing under the definitions "knowledge shared" and "knowledge received", and also take this approach to addresses the other dimensions of knowledge identified in this framework.

2.3 Innovation Capability

Literature on innovation capabilities has covered the following topics: [27] explored the impact of intellectual capital both in incremental and radical innovation. Among their findings, they verified a positive relationship between organizational capital and radical innovative capability. On the other hand, they noticed a negative relationship between human capital and radical innovative capability. [28] confirmed the influence of certain aspects of organizational learning orientation in firm's innovative capability, specifically: commitment to learning, shared vision, open-mindedness, and intra-organizational knowledge sharing. [29] here the authors link inter-firm relationship strengths and tacitness of knowledge transfer, its extent and innovation capability, and also innovation capability and innovation performance, all this based on the theory of knowledge.

2.4 Tacit Knowledge

Authors have focused in grasp the relevance of this kind of knowledge at the organizational level. [30] this paper summarized and analyzed the most widely spread techniques used in organization to acquire, measure, teach, share and apply knowledge, also presents methods to use both tacit and explicit knowledge more effectively in organizations. [31] focused on generate ways to assess and use tacit knowledge using Polanyi's theories. The author presented the way in which intranet documents can be used to make tacit knowledge tangible without become it explicit, been this one of the reasons why people would rather not to share tacit knowledge, specifically, the three reasons for this phenomena according stated in that study are: (1) People is not conscious about the tacit knowledge they poses, (2) for people's use is not necessary to make it explicit, and (3) share tacit knowledge might represent lose competitive advantage. [32] summarized some barriers that organizations might face to transform tacit knowledge into the explicit kind, also some recommendations are presented. [33] examined the relevance of tacit knowledge for the innovation process and show how different geographical location affects the dynamics between knowledge and innovative activity, and the extent to which this interaction influences the geography of innovation and economic activity.

2.5 Explicit Knowledge

This kind of knowledge it's easier to transmit among individuals and groups, in fact, one of the main objectives of organizations it's to recognize tacit knowledge and transform it into explicit, or at least be able to transform as much knowledge as possible, it's important to stress this point given that compelling argumentation included in previous literature, for instance [34] basically states that tacit knowledge cannot be transform into explicit. Also, more recently [35, 36] argued that there are certain dimensions of knowledge that can't be reach. Opposite to tacit knowledge which is personal in nature (know-how), explicit knowledge needs to be created and neutered on the basis of tacit knowledge [37]. [38] also distinguished two categories of explicit knowledge, which are: object-based and rule-based. Object-based explicit knowledge is fixed into artifacts and often is presented through symbols, while rule-based explicit knowledge tent to be represented rules or organization's procedures [39].

2.6 Knowledge Quality

The concept of knowledge quality has been operationalized containing accuracy, timeliness, and usefulness. [40] moreover, previous literature has linked quality of knowledge with superior levels of trust. [41] explained the value of accurate information and its effects on the perception of trustworthiness, focusing on the relations between managers and employees. The author argued that workers consider their managers as trustworthy when the information they share was accurate and forthcoming, plus, clear explanations and opportune feedback when taking decisions cause superior levels of trust. [42] in this vine of research, stated that when subordinates believe that the information they received from their superiors was accurate the perception of trust increased. The authors also probed that in the cases where subordinates trust their superiors both the desire for interaction and the satisfaction with the communication raised. On the other hand, low levels of trust were linked with blocking or withholding the upward information flow. Based on those previous findings, in the present research is presumed that the lack of quality information flow also cause similar effects in horizontal relationships among colleagues, and that the externalities implicit in Guanxi might be a major factor that block or diminish significantly the stream of quality information between all the members in organizations, hence contracting innovation capability of firms.

2.7 Knowledge Quantity

Knowledge quantity is regarded as the perception of members in the organization that they are getting enough information [43]. Literature focused on this concept includes [44], in that paper, they explored the frequency of information been shared among managers in a multinational company, and confirmed that the perception of trustworthiness of the trusty and the trustor are closely related with the frequency of information shared between parties. [45] focused on the assumption that more communication is better for organizations, and this is defined as a "communication metamyth". Among the most salient findings of this research is that employees from all the organizations studied expressed their yearning for more information coming from different sources, this is, formal and informal ones. Also, this article explained that desire for more information is based on the assumption that organizational problems can be elucidated through further communication.

[46] based on the Uncertainty Reduction Theory explains how the provision of additional information among staff members benefits organizations, specifically, by reducing uncertainty levels and at the same time spurring satisfaction with the communication process. That study also explains the extent to which audits can add to better communication strategies in organizations. Therefore, acknowledging the relevance that variations in frequency or amount of knowledge flow can generate in organizations, it's worthy to analyze the extent to which Guanxi as a moderator can affect the flow of knowledge sharing among staff members, and finally the overall impact in the innovation capability of firms.

3 Cultural Dissimilarities between China and Both Confucian and Non-Confucian Societies

Most of the previous research on knowledge sharing behaviors and its links to innovation had been conducted in western countries, for this reason the results and even the very approaches taken to conduct those might not fit in different cultural context or countries with rather opposite economic and political models such as China, even considering the results obtained in different organizational studies in other countries in northeast Asia might not apply to the Chinese framework since the social norms embedded in those societies involve different a set of values. Japan and Korea share roots with the Chinese culture and so the Confucian precepts are also embraced in those countries, never the less some substantial variations can be acknowledged. Different authors have identified Wa as the analogous social code of Guanxi applied in Japan. Wa can be understood as harmony, and it's composed by other social values that basically aim to foster harmony and unity in the Japanese society [47]. Based on this facts and previous studies on Guanxi, in which it has been considered rather as a dyadic or close-network figure in society [1] we can identify a key difference from Wa. This Japanese social norm aims to foster a collectivistic society placing the interest of the community or organization first instead of individuals, while the main objective of Guanxi relationships is to provide benefits and advantages for individuals instead of the whole community. Therefore, it could be assumed that the relationship between knowledge sharing behaviors and the innovation capability of the firms in a social context based on the values of Wa would be substantially different from results obtained in China.

Another country with strong bases in the Confucian theory is Korea, in which the philosophy of Inhwa is the cornerstone of the society arguably at the same extent that Guanxi is in China, although the concept of kibun it's very important in the Korean society as well [48]. Inhwa, is similar to the Japanese concept of Wa, in that, it foster the ideas of harmony in society, kibun can be interpreted as a set of tools to reach Inhwa. This social norm stresses the importance preserving pride, face, and a peaceful state of mind for the self as well as for others, thus Inhwa can be seen as more inclusive social norm as compare with Guanxi. For this reason, as previously mentioned in the case of Wa, the knowledge sharing behaviors of employees and its impact in the innovation capability of the firm under the set of values contained in Inhwa might present divers trends as compare with China.

The knowledge sharing behaviors and their impact in innovation capability of firms in the Chinese context it's also very likely to differ from non-Confucian-based societies. For instance, Wasta, which is an indigenous concept of Arab countries refers to a process whereby individuals can achieve goals by networking with key persons [14]. [49] Wasta relationships are often times a source of pride and prestige for the parties involved. While in China Guanxi relationships have a rather negative connotation in the society.

[14] explained that the similarities found between different social norms or indigenous influence styles across cultures does not diminish the importance of analyzing those separately in order to grasp particular organizational behaviors in those locations. Also, they argued that the breakthroughs got by studying each of those indigenous styles alone might be useful in different locations. Therefore, the results obtained in this research as well as the conclusions reached will not only be applicable in the Chinese context, but also have the potential to help researches and managers to understand the same phenomena in different cultural context.

Moreover, the impact of local cultures in knowledge sharing behaviors has been addressed in previous literature such as [50] they explained how collectivistic and individualistic cultures practice different methods to treat information and create knowledge. So it can be accepted that knowledge sharing behaviors, it's externalities as well as the solutions to those should not be analyzed using the "one-size-fits-all" approach, instead general theories of organizational behavior should be consider along with the due specific socio-cultural connotations.

4 Propositions

According with the analysis of previous literature conducted for the present study as well as the approach taken for it, a series of propositions are presented next:

Proposition 1: The correlation between type of knowledge shared and the innovation capability of firms will be moderated by Guanxi such that the correlation will be significantly less strong when a high level of Guanxi is verified.

Proposition 2: The correlation between type of knowledge received and the innovation capability of firms will be moderated by Guanxi such that the correlation will be significantly less strong when a high level of Guanxi is verified.

Proposition 3: The correlation between the quality of knowledge shared and the innovation capability of firms will be moderated by Guanxi such that the correlation will be significantly less strong when a high level of Guanxi is verified.

Proposition 4: The correlation between the quality of knowledge received and the innovation capability of firms will be moderated by Guanxi such that the correlation will be significantly less strong when a high level of Guanxi is verified.

Proposition 5: The correlation between the quantity of knowledge shared and the innovation capability of firms will be moderated by Guanxi such that the correlation will be significantly less strong when a high level of Guanxi is verified.

Proposition 6: The correlation between the quantity of knowledge received and the innovation capability of firms will be moderated by Guanxi such that the correlation will be significantly less strong when a high level of Guanxi is verified.

5 Economic Relevance

Concrete governmental actions to spur innovation in Chinese organizations had been taken by China's central government, which in 2005 released the Medium-and Long-term National Plan for Science and Technology

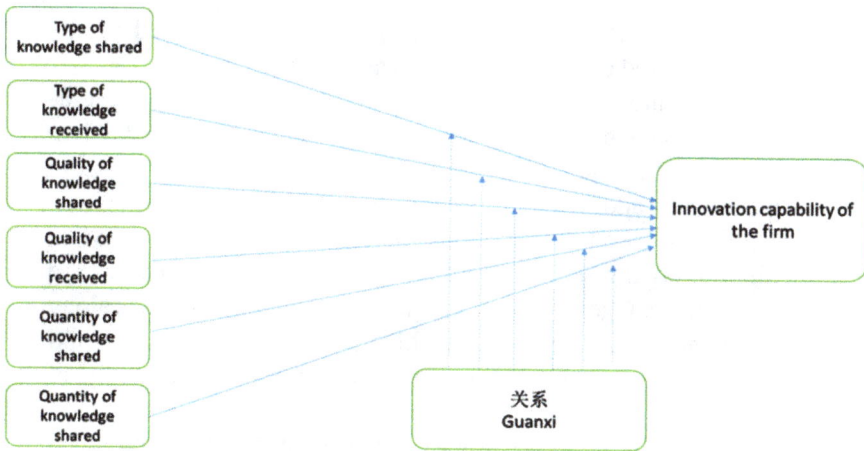

Figure 1   Diagram of the model proposed.

Development (2006–2020) [51]. It included the policy of 'indigenous innovation' as part of China's national strategy. [52] explored the evolution of the innovation process in China by identifying three stages of technology innovation: imitative innovation, cooperative innovation, and indigenous innovation. That study argued that even though China had achieve success in implementing the two previous stages of innovation (imitative and cooperative), indigenous innovation still had been elusive to Chinese organizations. Hence, the object of the present research is closely aligned with current managerial needs generated by the necessity to stimulate innovation capabilities pressing Chinese organizations.

6 Discussion

Both the theory of social capital and the indigenous social norm of Guanxi include precepts that stress the distinction between insiders and outsiders in relation with an individual's social circle. In this sense, we found the figures of private social capital in one hand and the relationalism embedded in Guanxi relationships in the other. By answering the main question of this research important contributions for the literature on organizational behavior could be obtained, as well as set the bases for future research. The main question to be addressed is:

Does the innovation capability of Chinese organizations is affected by the role of Guanxi as a moderator between the former and the knowledge sharing behaviors of their employees?

The main outcomes of conducting an empirical research based on the framework here presented can be summarized as follows:

First, identify the current state of Guanxi in Chinese organizations, specifically how it moderates the relationship between knowledge sharing behaviors among employees and the innovation capability of firms. The evolutive nature of this concept has been addressed in previous literature, for instance [53, 54]. Hence, this study is focused on determine the extent to which the ongoing changes in Chinese organizations generated by diverse factors, such as the generational shifts in management positions and new societal objectives along with a strong influence from the western world affects the relevance of Guanxi and its role between knowledge sharing and the innovation capability of firms [55].

Second, find the features of the Chinese organizations that have mitigated the externalities implicit in Guanxi, previous research points to organizational aspects that can influence knowledge sharing behaviors in despite of the presence of Guanxi, such as HR practices [2], also group size has been identified as a factor that can determine the practice of knowledge sharing behaviors [56]. The empirical development of the theoretical framework here presented also could shade light on the demographics in each organization and their role in the innovation capability, some individual characteristics to focus on could be age and gender given that their role in knowledge sharing dynamics has been proved [57, 58]. By classifying the relevant features of the organizations that have successfully reduced the externalities present in Guanxi, more organizations could consider, adapt and apply those managerial practices to deal better with indigenous social norms.

Third, present empirical bases to compare the role and evolution of Guanxi in intra-organizational knowledge sharing behaviors and innovation capabilities in Chinese organizations with the impact of analogue social norms at the organizational level in different countries, this in order to foster the design of more comprehensive and universal managerial principles.

7 Conclusion

To consider the particularities implicit in Guanxi at both individual and organizational level as well as verifying its role in the innovation capability of Chinese firms is a promising approach that addresses a core issue identified for China's central government and Chinese private companies. Also, comparing this Chinese social norm with analogue figures from both Confucian and non-Confucian societies as proposed in this theoretical framework would contribute to create more accurate managerial principals. The operationalization of the variables here presented as well as other aspects regarding methodology would be covered in a follow up work. An empirical study built on this conceptual framework could also include the analysis of second order variables that might explain different results across organizations such as: Industry, staff composition, age of the organization, etc. This would also contribute to identify more directions for new research.

References

[1] Chen, X.-P., and Chen, C. C. (2004). On the intricacies of the Chinese guanxi: a process model of Guanxi development. *Asia Pac. J. Manag.* 21, 305–324.

[2] Chen, C. C., and Chen, X.-P. (2008). Negative externalities of close Guanxi within organizations. *Asia Pac. J. Manage.* 26:3753.

[3] Batson, C. D., Batson, J. G., Todd, R. M., Brummett, B. H., Shaw, L. L., and Aldeguer, C. M. R. (1995) Empathy and the collective good: caring for one of the others in a social dilemma. *J. Pers. Soc. Psychol.* 68, 619–631.

[4] Chen, Y., Friedman, R., Yu, E., and Sun, F., (2009). Examining the positive and negative effects of Guanxi practices. *Asia Pac. J. Manag.* 28, 715–735.

[5] Liebowitz, J. (2002). Facilitating innovation through knowledge sharing: a look at the US Naval surface warfare center-cardrock division. *J. Comput. Inform. Syst.* 42.5, 1–6.

[6] Lin, H.-F. (2007). Knowledge sharing and firm innovation capability: an empirical study. *Int. J. Manpow.* 28, 315–332.

[7] Coleman, J. S. (1988) Social capital in the creation of human capital. *Suppl. Organ. Inst. Sociol. Econ. Approaches Anal. Soc. Struct.* 94, S95–S120.

[8] Van Buren, H., and Leana, C. R. (2000). "Building relational wealth through employment practices," in *Relational Wealth: The Advantages of Stability in a Changing Economy,* eds C. R. Leana and D. M. Rousseau (New York, NY: Oxford University Press), 233–246.

[9] Bourdieu, P., and Wacquant, L. J. D. (1992). *An Invitation to Reflexive Sociology,* Chicago, IL: University of Chicago Press.

[10] Wei, H., and Youmin, X. I. (2001). Is Guanxi a model of China's business? *Asia Pac. Manag. Rev.* 6, 295–304.

[11] Davies, H., Leung, T. K. P., Luk Yiu-hing Wong, S. T. K. (1995). The benefits of "Guanxi" the value of relationships in developing the Chinese market. *Ind. Market. Manag.* 24:207–214.

[12] Park, S. H., and Luo, Y. (2001). Guanxi and organizational dynamics: organizational networking in Chinese firms. *Strat. Manag. J.* 22, 455–477.

[13] Berger, R., Herstein, R., and Mitki, Y. (2013). Guanxi: the evolutionary process of management in China. *Int. J. Strat. Change Manag.* 5.

[14] Smith, P. B. (2011). Are indigenous approaches to achieving influence in business organizations distinctive? A comparative study of Guanxi, Wasta, Jeitinho, Svyazi and pulling strings. *Int. J. Hum. Resour. Manag.* 23, 1–16.

[15] Hutchings, K., and Weir, D. (2006). Guanxi and Wasta: a comparison. thunderbird international business review. *Thunderbird Int. Bus. Rev.* 48, 141–156.

[16] Huang, Q., Davison, R. M., and Gu, J. (2011). The impact of trust, Guanxi orientation and face on the intention of Chinese employees and managers to engage in peer-to-peer tacit and explicit knowledge sharing. *Inform. Syst. J.* 21, 557–577.

[17] Bala Ramasamya, Gohb, K. W., and Yeung, M. C. H.(2006). Is Guanxi (relationship) a bridge to knowledge transfer? *J. Bus. Res.* 59, 130–139.

[18] Buckley, P. J., Clegg, J., and Tan, H. (2006). Cultural awareness in knowledge transfer to China—The role of Guanxi and mianzi. *J. World Bus.* 41, 275–288.

[19] Fan, Y. (2007). Guanxi, government and corporate reputation in China: Lessons for international companies. *Market. Intell. Plann.* 25, 499–510.

[20] Ardichvili, A., Maurer, M., Li, W., Wentling, T., and Stuedemann, R. (2006). Cultural influences on knowledge sharing through online communities of practice. *J. Know. Manag.* 10, 94–107.

[21] Guthrie, D. (1999). *Dragon in a Three-Piece Suit: The Emergence of Capitalism in China.* Princeton, NJ: Princeton University Press.

[22] Yang, M. M. (2002). The resilience of Guanxi and its new deployments: a critique of some new Guanxi scholarship. *China Q.* 170, 459–476.

[23] Snejina Michailova, H. K. (2004). Facilitating knowledge sharing in Russian and Chinese subsidiaries: the role of personal networks and group membership. *J. Know. Manag.* 8, 84–94.

[24] van den Hooff, B., and de Ridder, J. A. (2004). Knowledge sharing in context: the influence of organizational commitment, communication climate and CMC use on knowledge sharing. *J. Know. Manag.* 8, 117–130.

[25] Ardichvili, A., Page, V., and Wentling, T. (2003) Motivation and barriers to participation in virtual knowledge-sharing communities of practice. *J. Know. Manag.* 7, 64–77.

[26] Subramaniam, M., and Youndt, M. A. (2005). The influence of intellectual capital on the types of innovative capability. *Acad. Manage. J.* 48, 450–463.

[27] Calantone, R. J., Tamer Cavusgila, S., and Zhao, Y. (2002). Learning orientation, firm innovation capability, and firm performance. *Ind. Mark. Manage.* 31, 515–524.

[28] Cavusgil, S. T., Calantone, R. J., and Zhao, Y. (2003). Tacit knowledge transfer and firm innovation capability. J. Bus. Ind. Mark. 18, 6–21.

[29] Smith, E. A. (2001). The role of tacit and explicit knowledge in the workplace. *J. Knowl. Manage.* 5, 311–321.

[30] Stenmark, D. (2001). Leveraging tacit organizational knowledge. *J. Manage. Inform. Syst.* 17, 9–24.

[31] Mahroeian, H., and Forozia, A. (2012). Challenges in managing tacit knowledge: a study on difficulties in diffusion of tacit knowledge in organizations. *Int. J. Bus. Soc. Sci.* 3:19.

[32] Howells, J. R. L. (2002). Tacit knowledge, innovation and economic geography. *Urban Stud.* 39, 871–884.

[33] Polanyi, M. (1962). *Personal Knowledge: Towards a Post-Critical Philosophy*. Chicago, IL: University of Chicago.

[34] Leonard, D. A., and Sensiper, S. (1998). The role of tacit knowledge in group innovation. *Calif. Manage. Rev.* 40, 112–132.

[35] Boisot, M. (2002). "The creation and sharing of knowledge," in *The Strategic Management of Intellectual Capital and Organizational Knowledge*, eds C. Choo and N. Bontis (New York, NY: Oxford University Press), 65–77.

[36] Choo, C. W. (1996). The knowing organization: how organizations use information to construct meaning, create knowledge and make decisions. *Int. J. Inform. Manage.* 16, 329–340.

[37] Nonaka, I., Toyama, R., and Byosière, P. (2001). "A theory of organizational knowledge creation: understanding the dynamic process of creating knowledge," in *Handbook of Organizational Learning and Knowledge*, eds M. Dierkes, A. Berthoin Antal, J. Child, and I. Nonaka (New York, NY: Oxford University Press), 491–516.

[38] Choo, C. W., and Bontis, N. (2002). *The Strategic Management of Intellectual Capital and Organizational Knowledge*. Oxford: Oxford University Press.

[39] Thomas, G. F., Zolin, R., and Hartman, J. L. (2009). The central role of communication in developing trust. *J. Bus. Commun.* 46, 287–310.

[40] Whitener, E. M., Brodt, S. E., Audrey Korsgaard, M., and Werner, J. M. (1998). Managers as initiators of trust: an exchange relationship framework for understanding managerial trustworthy behavior. Acad. Manage. Rev. 23, 513–530.

[41] Roberts, K. H., and O'really, C. A. (1974). Failures in upward communication in organizations: three possible culprits. *Acad. Manage. J.* 17, 205–215.

[42] Becerra, M., and Gupta, A. K. (2003). Perceived trustworthiness within the organization: the moderating impact of communication frequency on trustor and trustee effects. *Organ. Sci.* 14, 32–44.

[43] Haas, J. W., Sypher, B. D., and Zimmermann, S. (1996). A communication metamyth in the workplace: the assumption that more is better. *Int. J. Bus. Commun.* 33, 185–204.

[44] Hargie, O., Tourish, D., and Wilson, N. (2002). Communication audits and the effects of increased information: a follow-up study. *J. Bus. Commun.* 39:414–436.

[45] Alston, J. P. (1989). Wa, guanxi, and inhwa: managerial principles in Japan, China, and Korea. *Bus. Horiz.* 32, 26–31.

[46] Barnett, A., Yandle, B., and Naufal, G. (2013). *Regulation, Trust, and Cronyism in Middle Eastern Societies: The Simple Economics of 'Wasta.' Leibniz Information Centre for Economics, Discussion Paper No. 7201.* Available at: http://ssrn.com/abstract=2219126

[47] Bhagat, R. S., Kedia, B. L., Harveston, P. D., and Triandis, H. C. (2002). Cultural variations in the cross-border transfer of organizational knowledge: an integrative network. *Acad. Manage. Rev.* 27, 204–221.

[48] Medium- and long-term national plan for science and technology development (2006–2020). 国家中长 期科学和技术发展规划纲要 (2006–2020 年). Available at: http://www.gov.cn/gongbao/content/2006/content_240244.htm

[49] Ru, P., Zhi, Q., Zhang, F., Zhong, X., Li, J., and Su, J. (2012). Behind the development of technology: the transition of innovation modes in China's wind turbine manufacturing industry. *Energy Policy* 43, 58–69.

[50] Guthrie, D. (1999). *Dragon in a Three-Piece Suit: The Emergence of Capitalism in China*. Princeton, NJ: Princeton University Press.

[51] Yang, M. M.-H. (2002). The Resilience of guanxi and its new deployments: a critique of some new guanxi scholarship. *China Q.* 170, 459–476.

[52] Berger, R., Herstein, R., and Mitki, Y. (2013). Guanxi: the evolutionary process of management in China. *Int. J. Strateg. Change Manage.* 5, 30–40.

[53] Chase, R. (1998). The people factor. *People Manage.* 4:38.

[54] Sveiby, K.-E. (2003). *Personal Conversation.* Brisbane, QLD: Griffith Business School.

[55] Sawng, Y. W., Kim, S. H., and Han, H.-S. (2006). R&D group characteristics and knowledge management activities: a comparison between ventures and large firms.

[56] *Int. J. Technol. Manage.* 35, 241–261.

Biography

O. J. J. Torres is a senior Ph.D. candidate at the School of Management in Harbin Institute of Technology, in Harbin, PRC. His research is focused on the study and enhancement of the innovation capabilities of Chinese firms in the IT industry. He received an MBA degree from Beihua University in Jilin, China in 2012. In 2008 he earned his B.A. from Santa Maria University in Caracas, Venezuela.

Applying Total Productive Maintenance in Aluminium Conductor Stranding Process

O. Joochim[1] and J. Meekaew[2]

[1]*Institute of Field Robotics, King Mongkut's University of Technology Thonburi, Bangkok, Thailand*
[2]*Aluminium Conductor Department, Bangkok Cable Co., Ltd. (Chachoengsao Factory), Chachoengsao, Thailand*
E-mail: orapadee.joo@kmutt.ac.th; jumnong.m@bangkokcable.com

Received 20 July 2016; Accepted 10 September 2016
Publication 30 September 2016

Abstract

The purpose of this paper is to apply a total productive maintenance (TPM) technique for increasing the effectiveness in producing aluminium stranded conductors in order to reduce waste problems of the machine and improve the quality of the production. Two pillars of TPM activities for autonomous maintenance and focused improvement are established. Machines that have low overall equipment efficiency (OEE) are chosen as prototype machines. Training program for employees about the production work, cleaning activities to discover abnormal conditions, corrective actions during abnormal operations and a one point lesson (OPL) to educate operators for the production process are implemented. Autonomous maintenance standards are also created. Pareto analysis is used to quickly determine the critical equipment in the factory. A corrective action team is selected for improving the operation of the process. The research is evaluated by comparing the OEE of prototype machines based on production problems occurred before and after the improvement. The results show that the TPM implementation reduces the downtime from 7,730.80 minutes per month to 4,942.20 minutes per month, the loss of the scrap from 4,570.00 kilograms per month to 2,236.67 kilograms per month. The OEE is increased from 67.21 percent to 72.14 percent.

Journal of Industrial Engineering and Management Science, Vol. 1, 19–42.
doi: 10.13052/jiems2446-1822.2016.002

Keywords: Total productive maintenance, overall equipment efficiency, one point lesson.

1 Introduction

Thailand's residential electricity consumption continues to grow according to the economic growth of the country. The rapid increase in Thailand's electricity demand requires a substantial amount of electric wires and cables. Bangkok Cable Company Limited is a leader in manufacturing electric wires and cables in Thailand. Bangkok Cable's first factory is located in Samutprakan Province, and its second factory is situated in Chachoengsao Province. The company offers covering all types of bare and insulated conductors.

The electric cable production process is demonstrated in Figure 1. The production process begins with feeding aluminium rods into the melting process to produce aluminium conductors. In the next step, the conductors are extruded. After the extrusion, the conductors are stranded. The conductors are then packaged into the coils and reels. Insulated conductors must be made before packaging, if the conductors are required to be insulated or covered. Three primary types of insulating materials used are polyvinyl chloride (PVC), polyethylene (PE) and cross-linked polyethylene (XLPE). Production quality control is strictly executed on every step of production for the highest quality.

However, it was found from the production data collection of aluminium conductors between July to December 2014 that the overall equipment

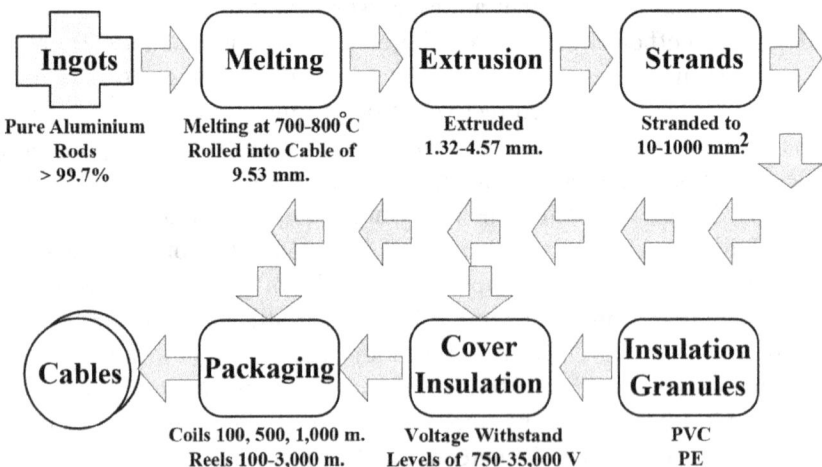

Ingots	Melting	Extrusion	Strands
Pure Aluminium Rods > 99.7%	Melting at 700-800°C Rolled into Cable of 9.53 mm.	Extruded 1.32-4.57 mm.	Stranded to 10-1000 mm².

Cables	Packaging	Cover Insulation	Insulation Granules
	Coils 100, 500, 1,000 m. Reels 100-3,000 m.	Voltage Withstand Levels of 750-35,000 V	PVC PE

Figure 1   Electric cable production process.

efficiency (OEE) is equal to 80.96 percent. OEE of the production of aluminium stranded conductors is equal to 70.46 percent which is the lowest comparing with other stages of the production. By increasing OEE, time and waste losses will be decreased. Therefore, it is possible to expand the production in the future. The aim of this paper is to study the effectiveness and implementation of TPM program for the production process of aluminium stranded conductors of an electric wire and cable manufacturer (i.e., Chachoengsao Factory of Bangkok Cable) [1].

2 Study Methodology

2.1 Total Productive Maintenance (TPM)

TPM has been acknowledged as a manufacturing strategy for helping to increase the productivity and overall equipment effectiveness. The concepts of TPM has been introduced and developed by in M/S Nippon Denso Co., Ltd of Japan in 1971. TPM has been widely implemented, and there are a number of case studies of TPM applications in the literature [2].

The basic practices of TPM are frequently called the pillars or elements of TPM. The core TPM is categorized into eight TPM pillars or activities

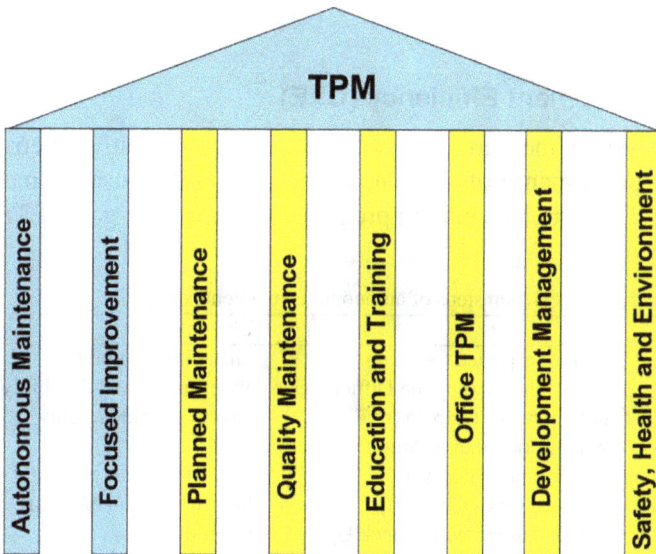

Figure 2 Eight pillars of TPM (from JIPM).

for accomplishing the manufacturing performance improvements. The eight pillars are autonomous maintenance, focused improvement (Kobetsu Kaizen), planned maintenance, quality maintenance, education and training, office TPM, development management, as well as safety, health and environment [3]. In order to increase the overall effectiveness and reduce times and wastes during the production of aluminium stranded conductors, autonomous maintenance and focused improvement which are the main TPM pillars are used in this paper.

Autonomous maintenance means maintaining one's own equipment in a good condition by one self. The purpose of this pillar is to prepare the operators to be able to take care of small maintenance tasks, hence allowing the skilled maintenance operators for spending time on more value added activity and technical repairs. The operators are therefore responsible for upkeep of their equipment on daily basis to prevent it from deteriorating or break down [3, 4]. There are 7 steps of autonomous maintenance as illustrated in Table 1.

Focused Improvement concentrates on reducing losses in the work place in order to improve operational efficiency. The targets of the improvement are zero losses (identify and eliminate losses), remove unsafe conditions, improve effectiveness of all equipment and reduce operation and maintenance costs. The principle of this pillar is that "a very large number of small improvement is more effective in an organization than a few improvement of large value" [4, 7].

2.2 Overall Equipment Efficiency (OEE)

OEE is an essential measure in TPM used as a quantitative metric for measuring the performance of a productive system. OEE is used to measure the success of TPM implementation program. The principle goal of TPM is

Table 1   Seven steps of autonomous maintenance [5, 6]

Change	Step	Objective
Machine Change	Step 1: Inspected Clean up	The ability to find the abnormality, and the ability to inspect the abnormality
	Step 2: Eliminating the Difficulties and Source of Outbreaks	
	Step 3: Formulating Autonomous Maintenance Standard	
Human Change	Step 4: Overall Check Up	Improvement of Machine Deterioration
	Step 5: Autonomous Check Up	
Environmental Change	Step 6: Standard Preparation	Maintenance Management from Machine Users
	Step 7: Continuous Improvement	

to increase the overall equipment efficiency. The three main components of OEE are equipment availability (A), performance efficiency (P), and quality rate (Q). OEE can be calculated as follows [8, 9].

$$OEE = A \times P \times Q \tag{1}$$

$$A = \left[\frac{Loading\ Time - Downtime}{Loading\ Time} \right] \times 100\% \tag{2}$$

$$P = \left[\frac{Ideal\ Cycle\ Time\ \times Total\ Pieces}{Operating\ Time} \right] \times 100\% \tag{3}$$

$$Q = \left[\frac{Total\ Product - Defects}{Total\ Product} \right] \tag{4}$$

2.3 One Point Lesson (OPL)

The OPL form is a learning tool which helps to communicate TPM training concepts to participants and employees. This form is structured for motivating the trainer to establish all substantial activities onto one simple and easy to use [6]. OPL is the lessons learnt by operators after carrying out the autonomous maintenance or focused improvement activities. A small-group-activity leader marks the activities into the OPL report. The enhancing of OPL allows the increase of the improvements done by the operators. Thus, the operators are becoming more skillful [10].

3 Case Study

As in Figure 3, The aluminium conductor stranding process starts with passing extruded aluminium conductors into each layer of the stranders. The conductors are stranded according to the standard of the strand length. Each conductor is pulled by using a capstan and the stranders are rotated at the same time so that the conductors are twisted. The aluminium stranded conductors are then sent to be stored.

3.1 Problem Definition

Table 2 shows the data of the production plans, good products, wastes, loading time and downtime of the production process for all departments were collected in 6 months (from July to December 2014). From Table 2, the

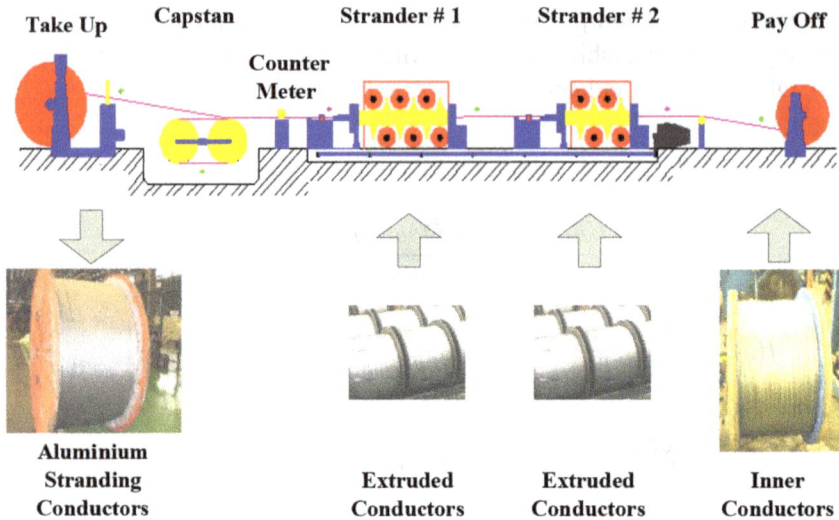

Figure 3   Electric cable production process.

Table 2   The production data of all departments

Department	Production Plans (Tons)	Good Products (Tons)	Wastes (Tons)	Loading Time (Hours)	Downtime (Hours)
Melting	6,202.08	6,132.77	69.31	1,838.84	140.94
Extrusion	5,533.32	5,530.85	2.47	6,178.94	1,014.93
Stranding	5,921.89	5,806.89	115.00	27,141.63	3,815.38
Medium Voltage Insulation	3,298.77	3,246.77	52.00	6,453.47	447.69
Low Voltage Insulation	5,206.89	5,169.65	37.24	12,672.84	1,175.80
PE and PVC Grains	3,794.84	3,793.50	1.34	7,559.00	285.00
Total	29,957.79	29,680.43	277.36	61,844.72	6,879.74

equipment availability (A), performance efficiency (P), quality rate (Q) and overall equipment efficiency (OEE) are calculated as in Table 3 [11].

The OEE of each department from Table 2 is shown in Figure 4. It can be noticed from Table 3 and Figure 4 that the average OEE score is equal to 80.96 percent. The OEE of the production of aluminium stranded conductors is equivalent to 70.46 percent which is the lowest comparing with other stages of the production. Therefore, this paper intends to increase the OEE of the stranding process. Table 4 demonstrates the production data of the aluminium conductor stranding process.

Table 3 Overall equipment efficiency of the production

Department	A (%)	P (%)	Q (%)	OEE (%)
Melting	92.34	95.90	98.88	87.56
Extrusion	83.57	93.63	99.96	78.22
Stranding	85.94	83.61	98.06	70.46
Medium Voltage Insulation	93.06	94.00	98.42	86.10
Low Voltage Insulation	90.72	93.00	99.28	83.77
PE and PVC Grains	96.23	96.50	99.96	92.83
Average	88.88	92.23	98.77	80.96

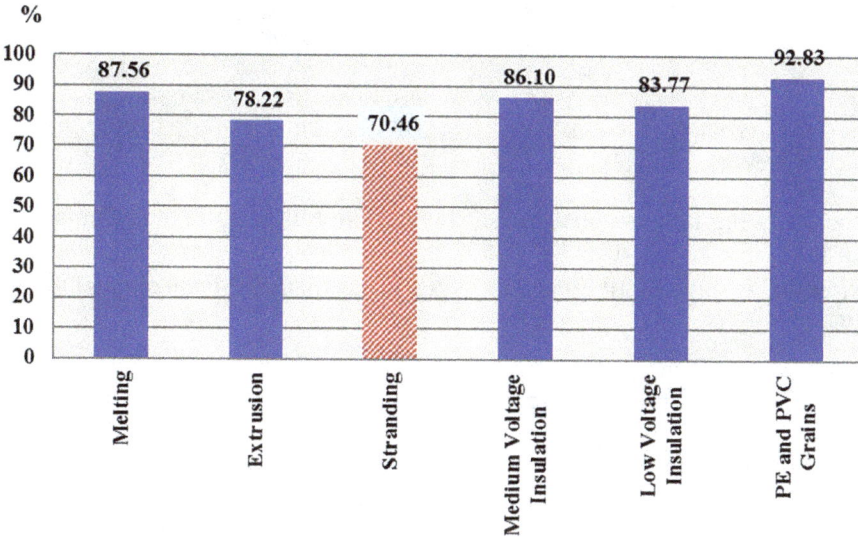

Figure 4 OEE of the production.

By using the data of Table 4, the sample of the OEE calculation for the ST09 strander machine is as follows.

$$A = \left[\frac{2{,}597.45\ hrs - 430.77\ hrs}{2597.45\ hrs} \right] \times 100\% = 83.42\%$$

$$P = \left[\frac{0.0011\ (hrs\ per\ m.) \times 2{,}400\ m.}{3.17\ hrs} \right] \times 100\%$$

P is calculated from each type of conductors before calculating for the average of the machine. For instance, the idle cycle time is equal to 0.001 hours per meter, the length of produced conductors is equivalent to 2,400 meters,

Table 4   Overall equipment efficiency of the production

Machine	Production Plans (Tons)	Good Products (Tons)	Wastes (Tons)	Loading Time (Hours)	Downtime (Hours)
ST01	288.62	276.12	12.50	2,879.46	353.76
ST04	635.27	617.87	17.40	2,808.54	462.38
ST06	558.47	533.72	24.75	3,097.18	646.67
ST08	196.22	192.09	4.13	2,327.04	366.65
ST09	561.07	535.75	25.32	2,597.45	430.77
ST11	273.04	254.54	18.50	2,690.00	440.75
ST02	205.20	204.80	0.40	2,298.25	155.62
ST03	1,392.00	1,388.60	3.40	2,976.81	306.27
ST05	1,560.55	1,553.35	7.20	3,188.47	521.00
ST13	251.45	250.05	1.40	2,278.43	131.50
Total	5,921.89	5,806.89	115.00	27,141.63	3,815.37

and the operating time is equal to 3.17 hours. The following is the calculation of P.

After the average of all conductor types is calculated, P is then equal to 76.83%.

$$Q = \left[\frac{561.07 \ tons \times 25.32 \ tons}{561.07 \ tons}\right] \times 100\% = 95.49\%$$

$$OEE = 0.832 \times 0.763 \times 0.9549 \times 100\% = 61.20\%$$

$$= 84.12\%$$

The OEE of strander machines are illustrated in Figure 5. The strander machines can be divided into two groups; (1) machines that loading the extruded conductors into steel wheels (i.e., ST01, ST04, ST06, ST08, ST09 and ST11), and (2) machines that loading the extruded conductors into baskets (ST02, ST03, ST05 and ST13). The ST03 and ST09 machines which have the lowest OEE compared with other machines of their group are selected as the prototype machines. The data of the problems of the downtime that do not include the preparation time of ST03 are collected.

The problems are arranged as demonstrated in Figures 6 and 8 depending on the priority of the downtimes and represented by the pareto graph for the data analysis. The collection of the data of quality loss problems is implemented. The arrangement of the problems is done according to the priority of the wastes. The data are illustrated as the pareto graph in Figures 7 and 9.

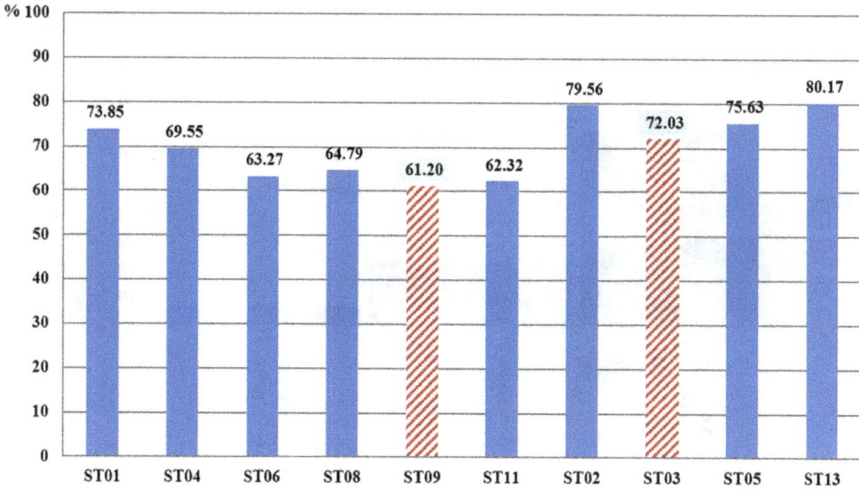

Figure 5   OEE of strander machines.

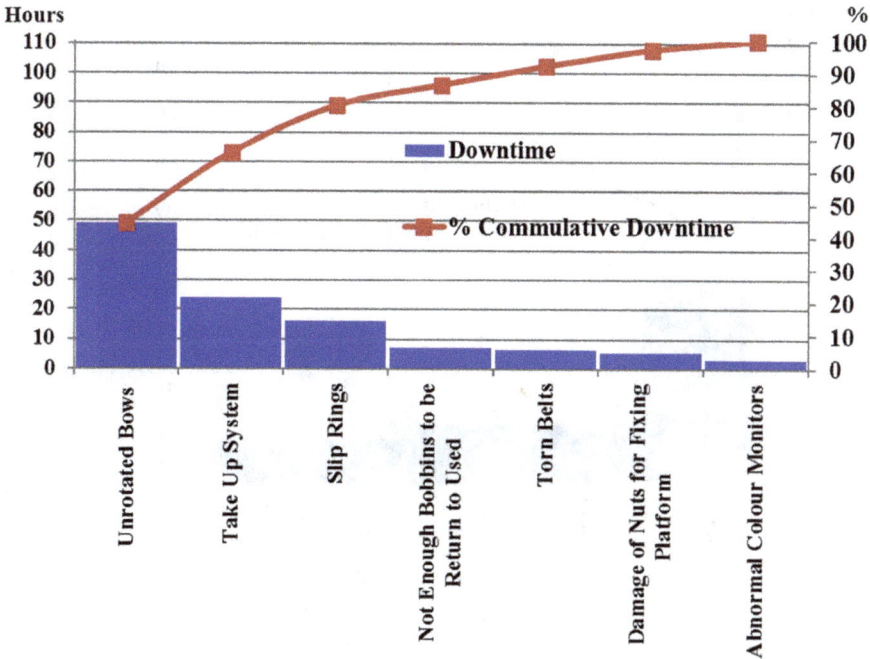

Figure 6   Pareto of the downtime problems of ST03 before the improvement.

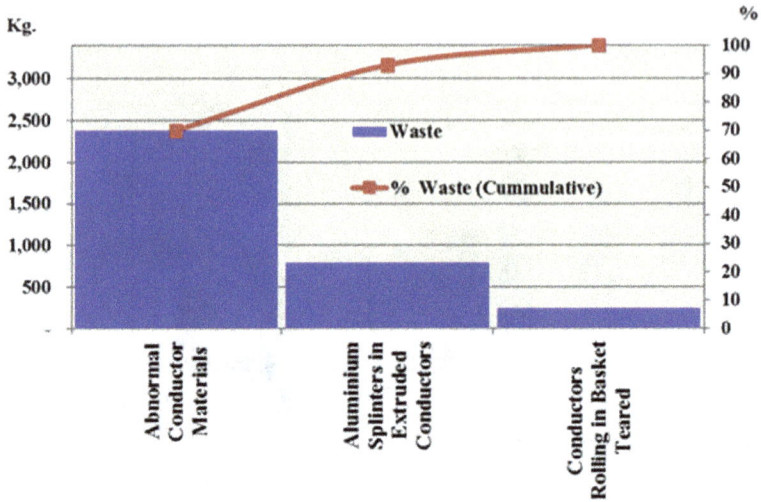

Figure 7   Pareto of the quality problems of ST03 before the improvement.

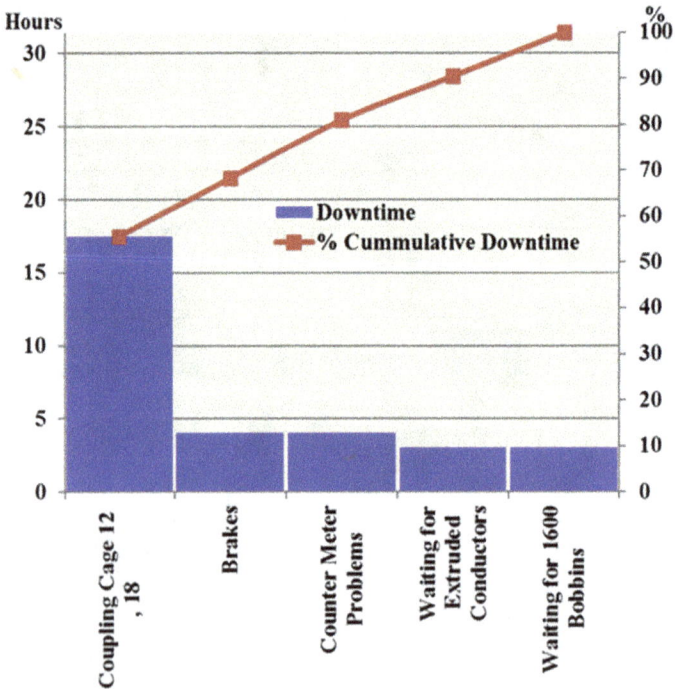

Figure 8   Pareto of the downtime problems of ST09 before the improvement.

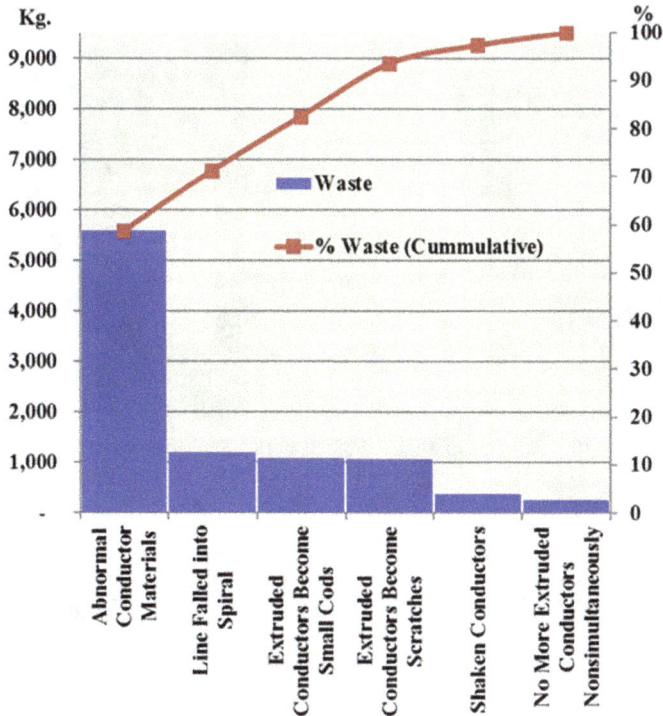

Figure 9 Pareto of the quality problems of ST09 before the improvement.

3.2 Implementation of Autonomous Maintenance

Autonomous maintenance is implemented by training the operators of the strander departments 1 and 2 about how to maintain the strander machines (self-maintenance). The operators are divided into two teams for the self-maintenance works for the ST03 and ST09 machines. The procedure begins by cleaning the machines, finding the abnormality and labeling to demonstrate the abnormality. The equipment used for the primary cleaning, protection and others is prepared. The maintenance is defined to do together every Wednesday.

The cleaning and finding of the abnormality are performed at the same time. The labels are attached at the abnormal parts (see Figure 10). The discovery dates and priorities are specified. The abnormality and correctional solution are defined. The list of the abnormality is made and recorded. The abnormality is corrected. The before and after maintenance data about the problems, solutions and benefits of the maintenance are list. The learning about the finding of the problems and solutions are shared to the related operators by

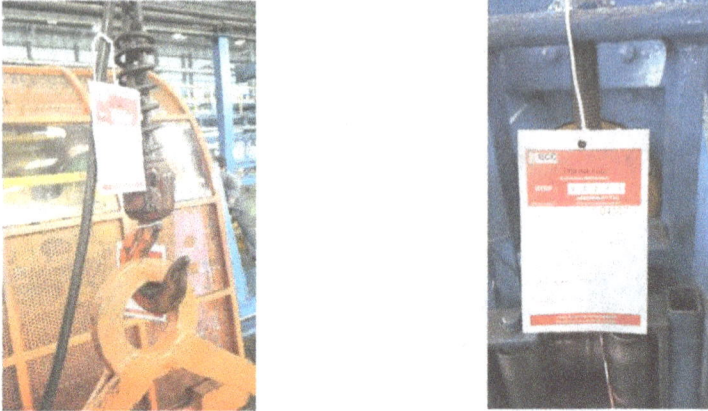

Figure 10 Samples of attached labels at the abnormal parts.

Table 5 Autonomous maintenance works

Procedures	Topics	Amount	Finished Solved	Operations (%)
1	Correction of Abnormality	60	53	88.33
2	Training Using OPL	25	20	80.00
3	Lubrication Standard	8	8	100.00
4	Self-Inspection Standard	4	3	75.00
	Total	97	84	86.60

using OPL. The training using OPL is classified according to the topic, basic knowledge, improvement work or problem. The area for the lubrication is defined. The standard of the lubrication is implemented. The self-inspection of the machines is performed. The self-inspection standard is created. The information about the machine components, lubrication standard, inspection of the operations of the equipment and observation method is used to create the inspection sheet for daily maintenance. Table 5 illustrates the amount of maintenance works. The improvements in abnormalities are listed. The photos of abnormalities before and after the improvement are compared as in Table 6.

3.3 Implementation of Focused Improvement

It can be seen from Figures 6 to 9 that a large majority of the problems (80 percent) are selected for the improvement according to the pareto analysis. The related workers are defined for the problems consisted of representatives

Table 6   Samples of abnormality before and after the improvement

No.	Before	After	Correction	Result
1			Remove unnecessary things	More clean and tidy
2			Repair the gate wheel	Wheel is not jerk, and rotates continuously.
3			Install the ring gear	The Gear is not remove.
4			Clean Grease	Equipment is clean, and easy for inspection.

from engineering, quality assurance and planning departments. The meeting is set for the related workers in order to solve the problems using the 5 W 1 H principle for inspecting the data of the problem characteristics.

The improvement is done corresponding to the defined plan. Subsequently, the downtimes and wastes are reduced. Focused Improvement starts with solving the problem of bows that cannot be rotated. The take up system doesn't work. The slip rings are electrocuted. The machines are shut down for fixing. The bows and slip rings are changed. The downtime problem is solved. During the stranding process, the conductors are teared. The cause of tearing is investigated, and the factors of tearing and standard of the inspection are defined. The frequency of inspection and responsible persons are specified. As a result, the losses are decreased. The problems of Coupling Cage 12 and 18 are occurred according to eroding of shafts. The shafts and couplings are changed. The data of the problems after the improvement are collected between January to March 2015, see Figures 11 to 14.

3.4 Production Data Before and After the Improvement

After the implementation of TPM for autonomous maintenance and focused improvement, the comparison between before the improvement (from October to December 2014) and after the improvement (from January to March 2015) are investigated. The selected prototype machines are ST03 and ST09 machines. The production data before and after the improvement of the ST03 machine are shown in Tables 7 and 8 respectively. According to Tables 7 and 8,

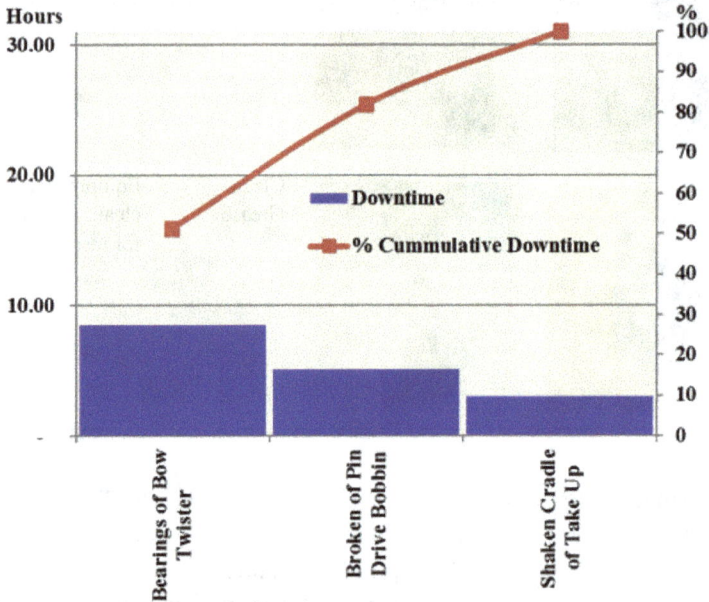

Figure 11   Pareto of the downtime problems of ST03 after the improvement.

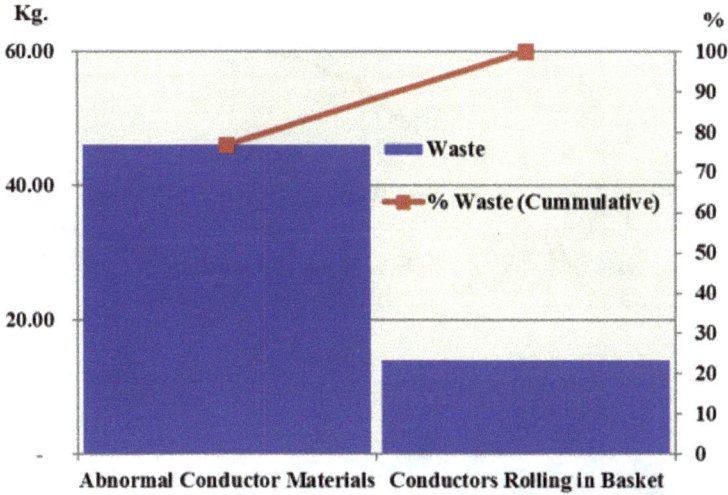

Figure 12 Pareto of the quality problems of ST03 after the improvement.

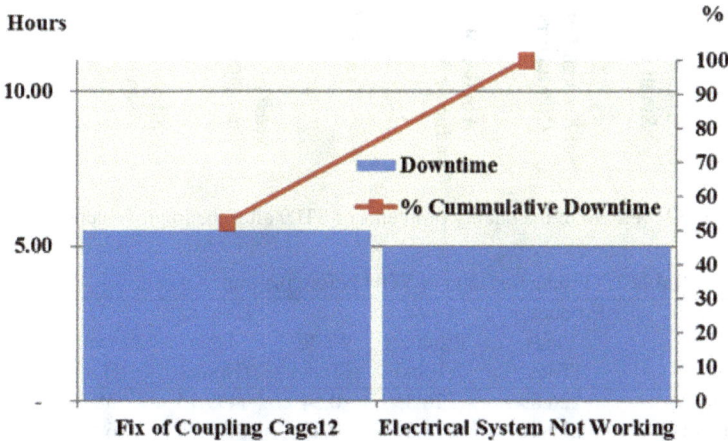

Figure 13 Pareto of the downtime problems of ST09 after the improvement.

the OEE before and after the improvement of ST03 are correspondingly calculated as shown in Tables 9 and 10.

It can be found from Tables 9 and 10 that the average per month of the equipment availability (A) is increased from 88.50 percent to 92.75 percent. The average per month of the performance efficiency is increased from 80.59 percent to 82.93 percent. The average per month of the quality rate is increased

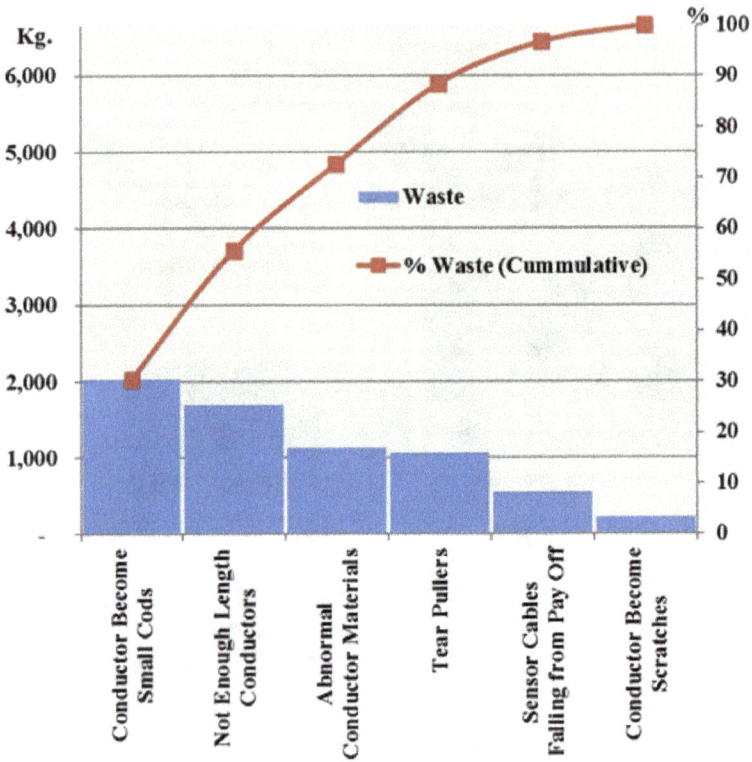

Figure 14   Pareto of the quality problems of ST09 after the improvement.

Table 7   Production data of ST03 before the improvement

Month (2014)	Production Plans (Tons)	Good Products (Tons)	Wastes (Tons)	Loading Time (Hours)	Downtime (Hours)
October	226.89	226.35	0.54	499.75	58.32
November	219.26	218.86	0.40	477.75	69.71
December	224.67	224.37	0.29	505.75	46.75
Total	670.81	669.58	1.24	1,483.25	174.78
Average/Month	223.60	223.19	0.41	494.42	58.26

from 99.82 percent to 99.99 percent. As a result, the average per month of OEE is absolutely increased from 70.97 percent to 76.89 percent.

The production data before and after the improvement of the ST09 machine are demonstrated in Tables 11 and 12 respectively. From Tables 11 and 12, the

Table 8   Production data of ST03 after the improvement

Month (2015)	Production Plans (Tons)	Good Products (Tons)	Wastes (Tons)	Loading Time (Hours)	Downtime (Hours)
January	249.59	249.57	0.02	567.00	40.20
February	79.44	79.42	0.018	225.75	16.16
March	149.77	149.75	0.02	356.75	27.00
Total	478.80	478.75	0.06	1,149.50	83.36
Average/Month	159.60	159.58	0.02	383.17	27.79

Table 9   OEE of ST03 before the improvement

Month (2014)	A (%)	P (%)	Q (%)	OEE (%)
October	88.33	82.78	99.76	72.95
November	85.41	76.98	99.82	65.63
December	90.76	82.00	99.87	74.32
Average/Month	88.22	80.59	99.82	70.97

Table 10   OEE of ST03 after the improvement

Month (2015)	A (%)	P (%)	Q (%)	OEE (%)
January	92.91	82.79	99.99	76.91
February	92.84	82.44	99.98	76.52
March	92.43	83.57	99.99	77.23
Average/Month	92.75	82.93	99.99	76.89

Table 11   Production data of ST09 before the improvement

Month (2014)	Production Plans (Tons)	Good Products (Tons)	Wastes (Tons)	Loading Time (Hours)	Downtime (Hours)
October	102.65	97.78	4.88	601.92	92.16
November	104.61	100.16	4.45	603.26	79.41
December	58.23	55.08	3.15	328.92	40.19
Total	265.49	253.01	12.47	1,534.10	211.76
Average/Month	88.50	84.34	4.16	511.37	70.59

OEE before and after the improvement of ST09 are respectively calculated in Tables 13 and 14.

From Tables 13 and 14, it can be seen that the average per month of the equipment availability (A) is increased from 86.20 percent to 84.85 percent. The average per month of the performance efficiency is increased from 77.22 percent to 80.10 percent. The average per month of the quality rate

Table 12   Production data of ST09 after the improvement

Month (2015)	Production Plans (Tons)	Good Products (Tons)	Wastes (Tons)	Loading Time (Hours)	Downtime (Hours)
January	26.67	26.45	0.22	208.50	28.25
February	260.29	258.12	2.18	544.75	85.50
March	213.54	209.29	4.26	327.50	50.00
Total	500.50	493.85	6.65	1,080.75	163.75
Average/Month	166.83	164.62	2.22	360.25	54.58

Table 13   OEE of ST09 before the improvement

Month (2014)	A (%)	P (%)	Q (%)	OEE (%)
October	84.69	76.60	95.25	61.79
November	86.84	75.32	95.75	62.63
December	87.78	79.75	94.59	66.22
Average/Month	86.20	77.22	95.30	63.44

Table 14   OEE of ST09 after the improvement

Month (2015)	A (%)	P (%)	Q (%)	OEE (%)
January	86.45	79.20	99.18	67.90
February	84.30	80.10	99.16	66.96
March	84.73	81.00	98.01	67.27
Average/Month	84.85	80.10	98.67	67.38

is increased from 95.30 percent to 98.67 percent. As a result, the average per month of OEE is absolutely increased from 63.44 percent to 67.38 percent.

The production data of downtime and wastes of ST03 before the improvement during October to December 2015 and after the improvement from January to March 2015 are compared as the bar graphs in Figures 15 and 16.

It can be noticed from Figures 15 and 16 that the downtime is reduced from 174.78 hours to 83.36 hours after the improvement, while the wastes are decreased from 1,240 to 60 kilograms.

The comparison as the bar graphs of the production data of downtime and wastes of ST09 before the improvement during October to December 2015 and after the improvement from January to March 2015 are illustrated in Figures 17 and 18.

It can be seen from Figures 17 and 18 that the downtime is reduced from 211.76 hours to 163.75 hours after the improvement, while the wastes are decreased from 12,470 to 6,650 kilograms.

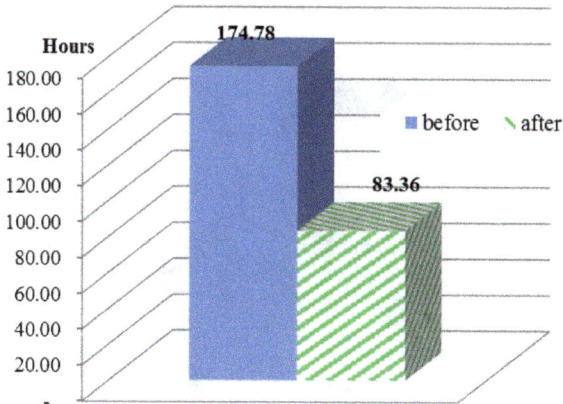

Figure 15 Downtime of ST03.

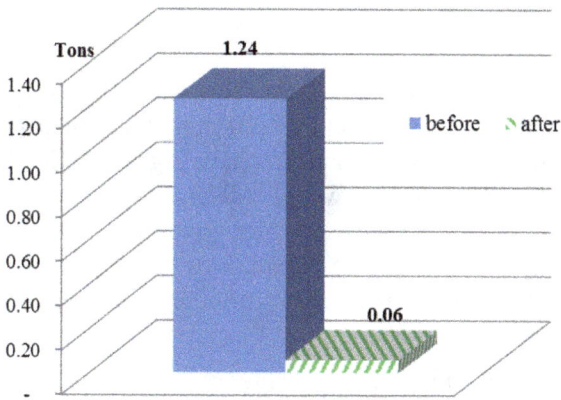

Figure 16 Wastes of ST03.

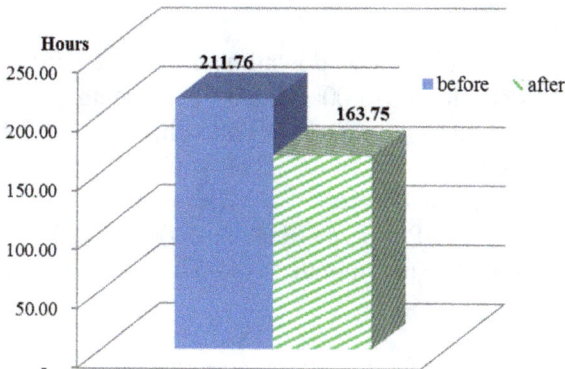

Figure 17 Downtime of ST09.

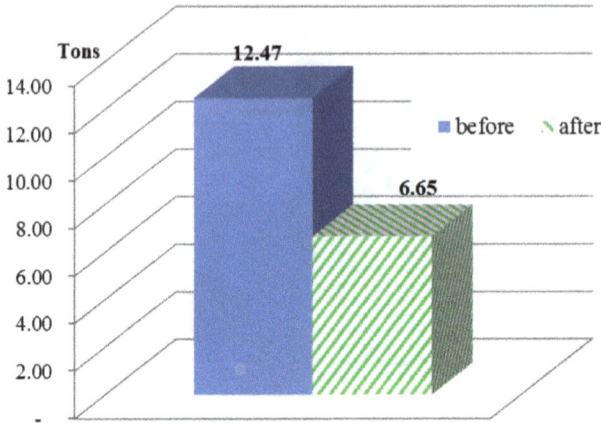

Figure 18   Wastes of ST09.

4 Concusion

The results from this paper show that TPM successfully gives the improvement. The wastes are decreased and the downtimes are also reduced. In addition, the aluminium stranded conductors have higher quality than before the improvement resulting in higher overall efficiency of the machines. These machines can be used to be the prototypes for operations of other machines of the company. Table 15 shows the comparison of OEE before and after the improvement. The results before and after the improvement can be summarized as below.

1) ST03 Prototype Machine: The downtime is reduced from 3,495.60 minutes per month to 1667.20 minutes per month. The waste is decreased from 413.33 kilograms per month to 20 kilograms per month. Before the improvement, OEE is 70.97 percent. OEE after the improvement is 76.89 percent.

2) ST09 Prototype Machine: The downtime is reduced from 4,235.20 minutes per month to 3,275.00 minutes per month. The waste is decreased from 4,156.67 kilograms per month to 2,216.67 kilograms per month. Before the improvement, OEE is 63.44 percent. OEE after the improvement is 67.38 percent.

3) Total of ST03 and ST09 Prototype Machines: The downtime is reduced from 7,730,80 minutes per month to 4,942.20 minutes per month.

The waste is decreased from 4,570.00 kilograms per month to 2,236.67 kilograms per month. Before the improvement, OEE is 67.21 percent. OEE after the improvement is 72.14 percent.

The following are suggestions for future research.

1) There are few operators for a big machine. Therefore, the extended period should be arranged for autonomous maintenance.
2) Other pillars of TPM should be activated.
3) Employment incentives should be offered to activate employees to work.
4) Due to slow collection of abnormality, the additional department should be offered for supporting the collection activities.

Table 15 Comparison of OEE before and after the improvement

Prototype Machine	Indicator	Unit	Before	After	Difference	
ST03	Downtime	minute/month	3,495.60	1,667.20	decrease	1,828.40
	Wastes	kg/month	413.33	20.00	decrease	393.33
	Equipment Availability (A)	percent	88.22	92.75	increase	4.53
	Performance Efficiency (P)	percent	80.59	82.93	increase	2.34
	Quality Rate (Q)	percent	99.82	99.99	increase	0.17
	Overall Equipment Efficiency (OEE)	percent	70.97	76.89	increase	5.92
ST09	Downtime	minute/month	4,235.20	3,275.00	decrease	960.20
	Wastes	kg/month	4,156.67	2,216.67	decrease	1,940.00
	Equipment Availability (A)	percent	86.20	84.85	decrease	1.35
	Performance Efficiency (P)	percent	77.22	80.10	increase	2.88
	Quality Rate (Q)	percent	95.30	98.67	increase	3.37
	Overall Equipment Efficiency (OEE)	percent	63.44	67.38	increase	3.94
Total	Downtime	minute/month	7,730.80	4,942.20	decrease	2,788.60
	Wastes	kg/month	4,570.00	2,236.67	decrease	2,333.33
	Equipment Availability (A)	percent	87.21	88.80	increase	1.59
	Performance Efficiency (P)	percent	78.91	81.52	increase	2.61
	Quality Rate (Q)	percent	97.56	99.33	increase	1.77
	Overall Equipment Efficiency (OEE)	percent	67.21	72.14	increase	4.93

References

[1] Joochim, O. and Meekaew, J. (2016). "Application of TPM in Production Process of Aluminium Stranded Conductors," in *Proceedings of International Conference on Industrial Engineering, Management Science and Application*, Jeju.

[2] Sethia, C. S., Shende, P. N., Dange, S. S. (2014). Total productive maintenance-a systematic review. *Int. J. Sci. Res. Dev.* 2, 124–127.

[3] Sütőová, A., Markulik, Š., Šolc, M. (2012). "Kobetsu Kaizen: its value and application," *Proceedings of Electronics International Interdisciplinary Conference*, 108–110.

[4] Gulati, R. (2013). *Maintenance and Reliability Best Practice*. 2nd edn. New York, NY: Industrial Press Inc.

[5] Sethia, C. S., Shende, P. N., and Dange, S. S. (2014). A case study on total productive maintenance in rolling mill. *J. Emerg. Technol. Innov. Res.* 1, 283–289.

[6] Hegde, H. G., Mahesh, N. S., and Doss, K. (2009). Overall Equipment Effectiveness Improvement by TPM and 5S Techniques in a CNC Machine Shop. *SASTECH* 8, 25–32.

[7] Paropate, R. V., and Sambh, R. U. (2013). The implementation and evaluation of total productive maintenance: a case study of midsized Indian enterprise. *Int. J. Appl. Innovat. Eng. Manage.* 2, 120–125.

[8] Siong, S. S., and Ahmed, S. (2007). *"TPM Implementation Can Promote Development of TQM Culture: Experience from a Case Study in a Malaysian Manufacturing Plant,"* in Proceedings of International Conference on Mechanical Engineering, Dhaka, Bangladesh.

[9] Meekaew, J. (2015). *Application of TPM for Productivity Improvement in Aluminium Conductor Stranding Production Process: A Case Study of Bangkok Cable Co., Ltd*. Thonburi: King Mongkut's University of Technology.

[10] Nakajima, S. (1988). *TPM Development Program*. Cambridge: Productivity Press.

[11] Aumor, T. (2004). *Total Preventive Maintenance*, 2nd edn. Bangkok: Thailand Productivity Institute, 113–148.

Biographies

O. Joochim received her Bachelor and Master of Engineering in Electrical Engineering from Assumption University and King Mongkut's Institute of Technology Ladkrabang, Bangkok, Thailand, respectively. She received her Doctoral Degree in Economics and Management from Leibniz University Hannover, Germany. She is currently a lecturer at Institute of Field Robotics, King Mongkut's University of Technology Thonburi, Thailand. Before she has joined with Institute of Field Robotics, she was respectively with Partnership Management & Project Follow-up and Vendor Performance Management Departments, Advanced Info Service PLC., Thailand.

J. Meekaew received his Bachelor of Science in Technical Education (Electrical Power Engineering) from Rajamangala Institute of Technology Thewet, Bangkok, Thailand. He received his Master of Science in Technopreneurship from Institute of Field Robotics, King Mongkut's University of Technology Thonburi, Thailand. At present, he works with Bangkok Cable Co., Ltd. (Chachoengsao Factory), Thailand.

Software Reliability Model Selection Based on Deep Learning with Application to the Optimal Release Problem

Yoshinobu Tamura[1] and Shigeru Yamada[2]

[1]Information Science and Engineering Section, Department of Engineering,
Graduate School of Sciences and Technology for Innovation, Yamaguchi University,
Ube-shi, Japan
[2]Department of Social Management Engineering, Graduate School of Engineering,
Tottori University, Tottori-shi, Japan
E-mail: tamura@yamaguchi-u.ac.jp; yamada@sse.tottori-u.ac.jp

Received 1 September 2016; Accepted 10 October 2016
Publication 21 October 2016

Abstract

In the past, many software reliability models have been proposed by several researchers. Also, several model selection criteria such as Akaike's information criterion, mean square errors, predicted relative error and so on, have been used for the selection of optimal software reliability models. These assessment criteria can be useful for the software managers to assess the past trend of fault data. However, it is very important to assess the prediction accuracy of model after the end of fault data observation in the actual software project. In this paper, we propose a method of optimal software reliability model selection based on the deep learning. Moreover, we show several numerical examples of software reliability assessment in the actual software projects. In particular, we discuss the optimal release time and total expected software cost in terms of the model selection based on the deep learning.

Keywords: Software reliability model, optimal model selection, deep learning, optimal release time, software cost.

Journal of Industrial Engineering and Management Science, Vol. 1, 43–58.
doi: 10.13052/jiems2446-1822.2016.003

1 Introduction

In actual software development projects, the comprehensive use of the technologies and methodologies in software engineering is needed for improving software quality/reliability, i.e., requirement specification, design, coding, and, testing. In particular, the waterfall model is well known as the sequential phases in software development process. In the past, many software reliability models [1–3] have been applied to assess the reliability for quality management and testing-progress control of software development. However, it is difficult for the software managers to select the optimal software reliability model for the actual software development project. As an example, the software managers can assess the software reliability for the past data sets by using usual model evaluation criteria. On the other hand, the estimation results based on the past fault data cannot be guaranteed for the future data sets of actual software projects.

The selection method of the optimal software reliability model based on the deep learning is proposed in this paper. Also, several numerical examples of software reliability assessment by using the fault data in the actual software projects are shown. Moreover, we compare the methods to estimate the cumulative numbers of detected faults based on the deep learning with that based on neural network. Furthermore, we discuss the optimal release time and total expected software cost in terms of the model selection.

2 Software Reliability Growth Models

Many software reliability models have been applied to assess the reliability for quality management and testing-progress control of software development. As an example, we show the typical software reliability models with the mean value function representing the expected number of detected faults during $(0, t)$ as follows:

Exponential NHPP (nonhomogeneous Poisson process) model

$$E(t) = a(1 - e^{-bt}),\tag{1}$$

Delayed S-shaped NHPP model

$$S(t) = a\{1 - (1 + bt)e^{-bt}\},\tag{2}$$

Logarithmic Poisson execution time model

$$\mu(t) = \frac{1}{\theta} \ln[\lambda_0 \theta t + 1],$$

(3)

Exponential SDE (stochastic differential equation) model

$$SDE_e(t) = a \left\{ 1 - e^{-bt + \frac{\sigma^2}{2}t} \right\},$$

(4)

S-shaped SDE model

$$SDE_s(t) = a \left\{ 1 - (1 + bt) e^{-bt + \frac{\sigma^2}{2}t} \right\},$$

(5)

where a is the expected number of initial inherent faults, b a fault-detection rate per unit time per fault, λ_0 the intensity of initial inherent failure, θ the reduction rate of the failure intensity rate per inherent fault, and σ a positive constant representing a magnitude of the irregular fluctuation. In this paper, we discuss above software reliability growth models.

3 Optimal Software Release Problem

Several optimal software release problems considering software development process have been proposed by many researchers [4, 5]. It is interesting for software developers to predict and estimate the time when we should stop testing in order to develop a highly reliable software system efficiently.

We define the following:

c_1: the fixing cost per fault during the testing phase,
c_2: the cost per unit time during the testing phase,
c_3: the maintenance cost per fault after the testing phase.

Then, the expected software cost in the testing phase can be formulated as:

$$C_1(t) = c_1 H(t) + c_2 t,$$

(6)

where $H(t)$ means the mean value function of software reliability growth model.

Also, the expected maintenance cost after the release of software is represented as follows:

$$C_2(t) = c_3 \{a - H(t)\}.$$

(7)

Consequently, from Equations (6) and (7), the total expected software cost is given by

$$C(t) = C_1(t) + C_2(t). \tag{8}$$

The optimum release time t^* is obtained by minimizing $C(t)$ in Equation (8).

4 Optimal Software Reliability Model Selection Based on Neural Network

Let $w_{ij}^1(i = 1, 2, \ldots, I; j = 1, 2, \ldots, J)$ be the connection weights from i-th unit on the sensory layer to j-th unit on the association layer, and $w_{jk}^2(j = 1, 2, \ldots, J; k = 1, 2, \ldots, K)$ denote the connection weights from j-th unit on the association layer to k-th unit on the response layer. Moreover, $x_i(i = 1, 2, \ldots, I)$ represent the normalized input values of i-th unit on the sensory layer, and $y_k(k = 1, 2, \ldots, K)$ are the output values. We apply the actual number of detected faults per unit time $N_i(i = 1, 2, \ldots, I)$ to the input values $x_i(i = 1, 2, \ldots, I)$.

Considering the amount of characteristics for the software reliability models, we apply the following amount of information as parameters $\lambda_i(i = 1, 2, \ldots, I)$ to the input data $x_i(i = 1, 2, \ldots, I)$.

- The estimated values of Akaike's information criterion
- The estimated values of mean square errors
- The estimate of all parameters included in model
- The estimated error between the estimated and actual values of the total number of detected faults in the testing time point of 25%, 50%, and 75% for data sets

The input-output rules of each unit on each layer are given by

$$h_j = f\left(\sum_{i=1}^{I} w_{ij}^1 x_i\right), \tag{9}$$

$$y_k = f\left(\sum_{j=1}^{J} w_{jk}^2 h_j\right), \tag{10}$$

where a logistic activation function $f(\cdot)$ which is widely-known as a sigmoid function given by the following equation:

$$f(x) = \frac{1}{1 + e^{-\theta x}}, \tag{11}$$

where θ is the gain of sigmoid function. We apply the multi-layered neural networks by back-propagation in order to learn the interaction among input and output [6]. We define the error function given by the following equation:

$$E = \frac{1}{2} \sum_{k=1}^{K} (y_k - d_k)^2,$$ (12)

where $d_k(k = 1, 2, \ldots, K)$ are the target input values for the output values. We apply 5 kinds of model type to the target input values $d_k(k = 1, 2, \ldots, 5)$ for the output values, i.e., the exponential NHPP model, the delayed S-shaped NHPP model, the Logarithmic Poisson execution time model, the exponential SDE model, and the S-shaped SDE model, respectively. Then, the number of units I in sensory layer is 37 because of 5 models.

5 Optimal Software Reliability Model Selection Based on Deep Learning

The structure of the deep learning in this paper is shown in Figure 1. In Figure 1, $z_l(l = 1, 2, \ldots, L)$ and $z_m(m = 1, 2, \ldots, M)$ mean the pre-training units. Also, $o_n(n = 1, 2, \ldots, N)$ is the amount of compressed characteristics. Several algorithms in terms of deep learning have been proposed [7–12]. In this paper, we apply the deep neural network to learn the software testing phase.

As with the neural network, we apply the 4 characteristic quantities of pre-training units discussed above. Moreover, we apply 5 kinds of model type to the amount of compressed characteristics, i.e., the exponential NHPP model, the delayed S-shaped NHPP model, the Logarithmic Poisson execution time model, the exponential SDE model, and the S-shaped SDE model, respectively.

6 Numerical Examples

We focus on actual software project data sets [13] in order to assess the performance of our method.

6.1 Estimation Results Based on Trend Analysis for Each Model

We apply 5 models as the typical two kinds of curve types, i.e., exponential and S-shaped reliability growth model curves for the total number of detected

[Pre-Training Units]
1st
Input and Output
Layer

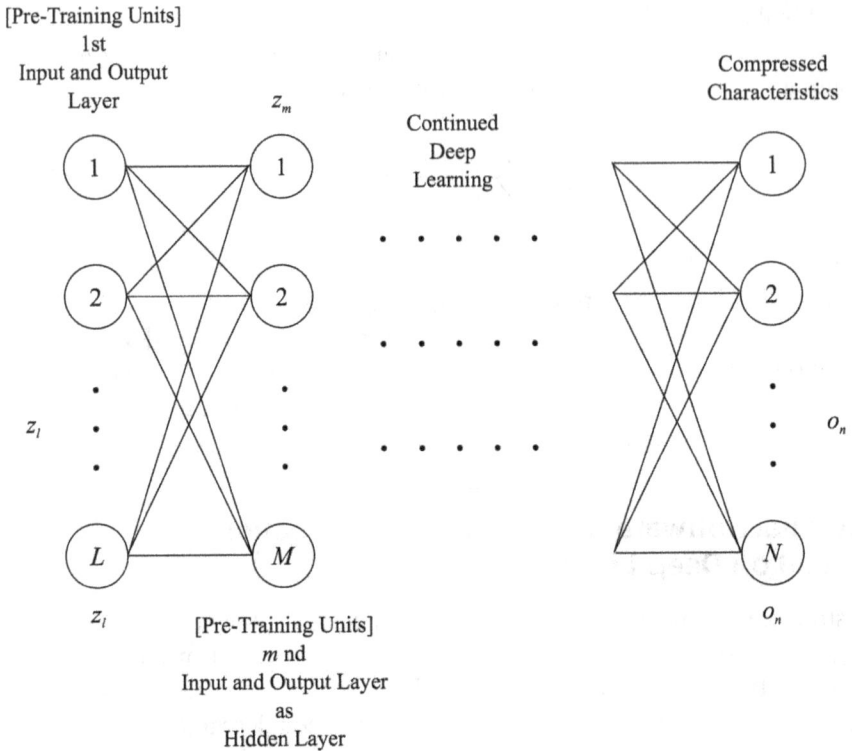

Figure 1   The structure of deep learning.

faults, to the actual fault data sets. The estimation results by using 90% of data set for each software development project are shown in Figures 2–4. Similarly, the estimation results by using 80% of data set for each software development project are shown in Figures 5–7. From Figures 2–7, we can confirm that the model best fitted for the past data are not always fitted for the future in fact. Then, it is necessary for the software managers to offer the model best fitted for the future.

6.2 Estimated Results of the Optimal Model Based on the Deep Learning

The estimated results of the optimal model based on the neural network and deep learning are shown in Tables 1 and 2. From Tables 1 and 2, we found that the estimated recognition rates based on the deep learning perform better than that of the neural network. In particular, the estimation results based on the deep learning give a high-recognition rates in terms of the type of model. Moreover, the estimation results based on the deep learning for the long-term prediction by using 80% of data sets perform significantly better than that of neural network.

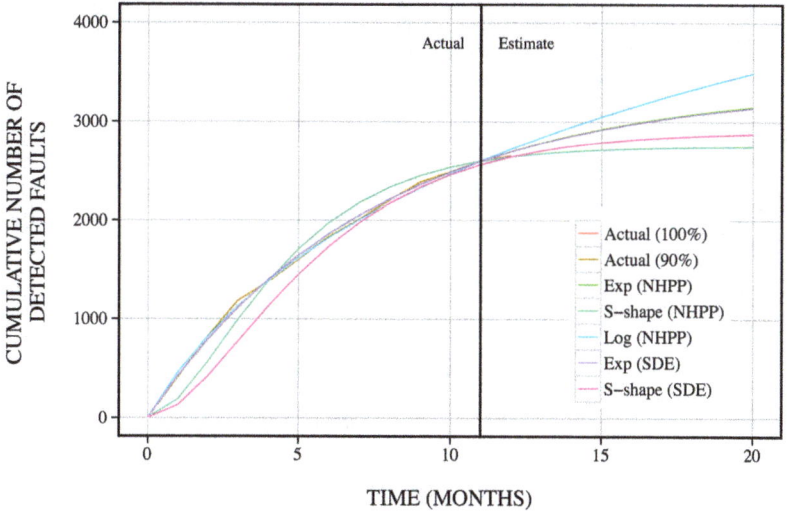

Figure 2 The estimation results by using 90% of data set of software project 1.

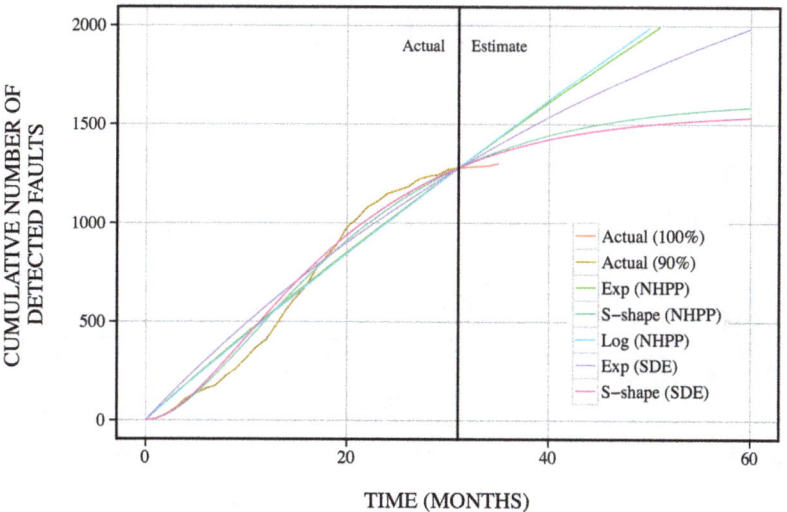

Figure 3 The estimation results by using 90% of data set of software project 2.

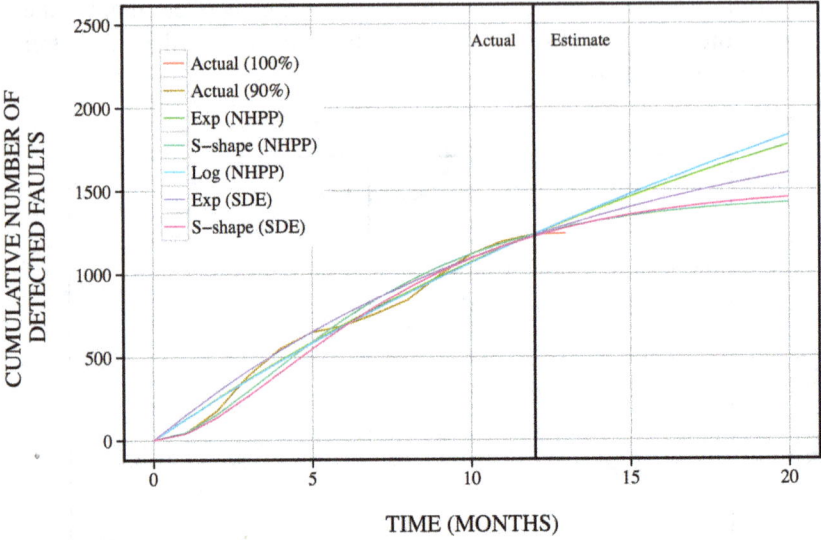

Figure 4   The estimation results by using 90% of data set of software project 3.

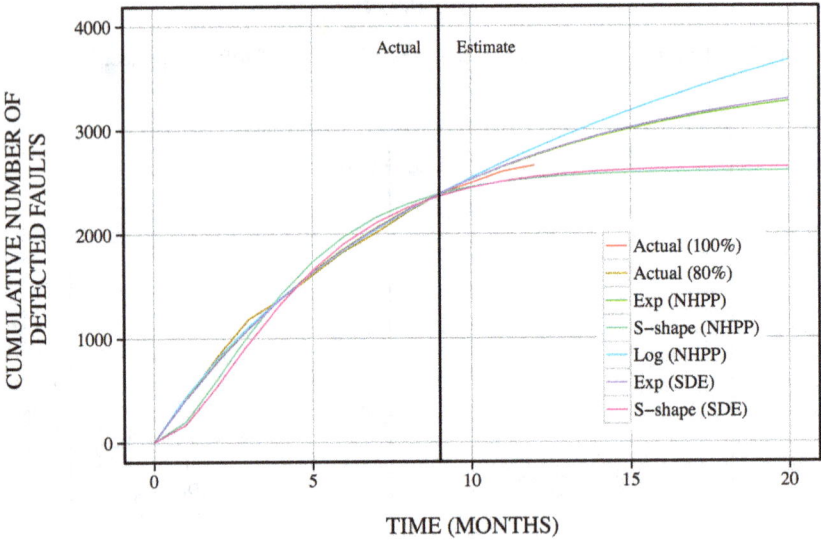

Figure 5   The estimation results by using 80% of data set of software project 1.

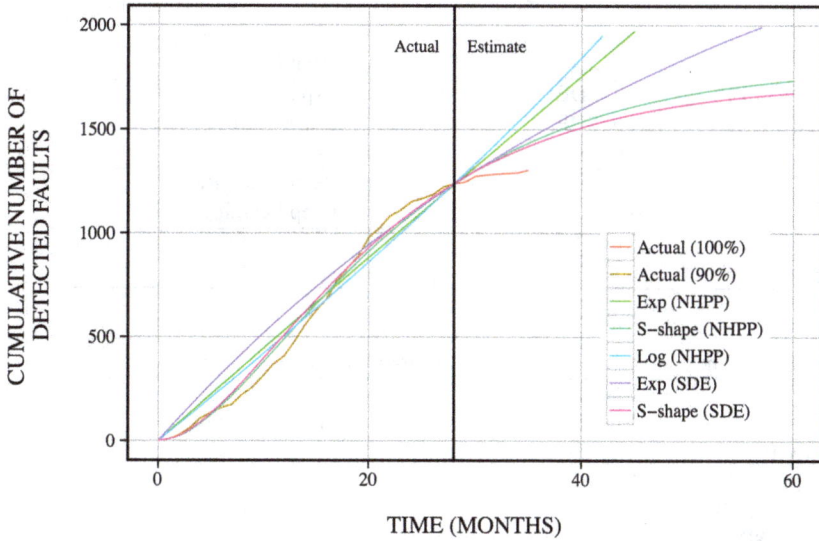

Figure 6   The estimation results by using 80% of data set of software project 2.

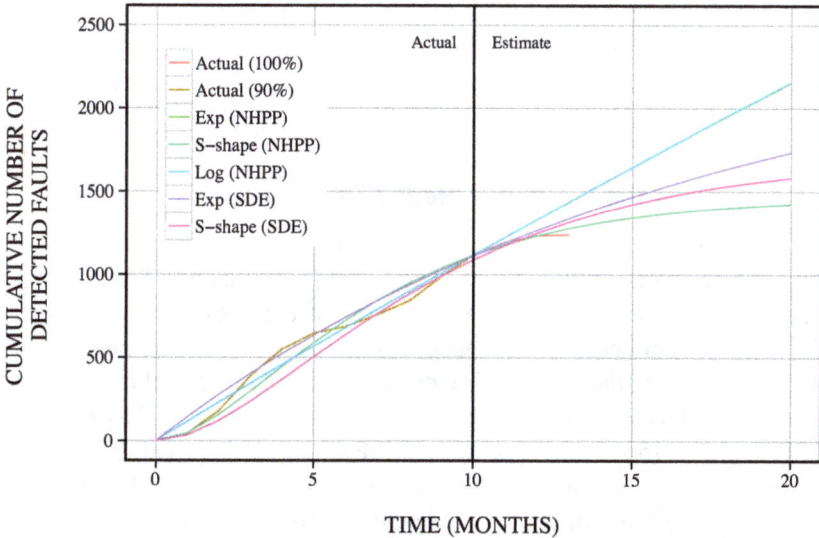

Figure 7   The estimation results by using 80% of data set of software project 3.

Table 1   The comparison of prediction accuracy after 90% of actual fault data

Accurate Model	Model Selected by Neural Network	Model Selected by Deep Learning
S-shape (NHPP)	Exp (SDE)	Exp (NHPP)
S-shape (SDE)	Exp (SDE)	S-shape (SDE)
S-shape (SDE)	Exp (SDE)	S-shape (NHPP)
	Recognition Rate by Neural Network	Recognition Rate by Deep Learning
Type of Model	0%	67%
Name of Model	0%	33%

Table 2   The comparison of prediction accuracy after 80% of actual fault data

Accurate Model	Model Selected by Neural Network	Model Selected by Deep Learning
Exp (NHPP)	Exp (SDE)	Log/Exp (NHPP)
S-shape (NHPP)	Exp (SDE)	Exp (SDE)
S-shape (SDE)	Exp (SDE)	S-shape (SDE)
S-shape (SDE)	S-shape (SDE)	S-shape (SDE)
S-shape (NHPP)	S-shape (SDE)	S-shape (SDE)
S-shape (SDE)	S-shape (SDE)	S-shape (SDE)
	Recognition Rate by Neural Network	Recognition Rate by Deep Learning
Type of Model	67%	83%
Name of Model	33%	50%

6.3 Discussion of the Optimal Release Time

We show the comparison results of the estimated optimal software release time and total expected software cost for each software project data in Tables 3–5, respectively. Then, we focus on 4 models without the logarithmic Poisson execution time model because this model assumes the initial inherent faults of infinity. Considering the prediction accuracy after 90% of actual fault data, the best fitted models are "S-shape (NHPP)" and "S-shape (SDE)" as shown in Table 1. Then, we found that "S-shape (NHPP)" and "S-shape (SDE)" is the shortest period of release time and the minimal cost from Tables 3–5. From these results, it is very important to decide the best model in terms of optimal release time, because the model selection has a great influence on the optimal release time and total software cost.

Table 3 The comparison of the estimated optimal software release time and total expected software cost in case of software project 1

Models	Optimal Release Time	Total Expected Software Cost
Exp (NHPP)	46.070	6827.9
S-shape (NHPP)	22.112	5528.0
Exp (SDE)	45.642	6791.7
S-shape (SDE)	27.040	5832.1

Table 4 The comparison of the estimated optimal software release time and total expected software cost in case of software project 2

Models	Optimal Release Time	Total Expected Software Cost
Exp (NHPP)	678.35	16872.7
S-shape (NHPP)	74.124	3331.3
Exp (SDE)	229.46	6429.4
S-shape (SDE)	68.989	3193.9

Table 5 The comparison of the estimated optimal software release time and total expected software cost in case of software project 3

Models	Optimal Release Time	Total Expected Software Cost
Exp (NHPP)	116.99	6415.2
S-shape (NHPP)	29.124	2951.2
Exp (SDE)	65.768	4155.1
S-shape (SDE)	31.475	3051.7

7 Conclusion

In the past, many software reliability models have been proposed. In fact, these software reliability models have been applied to many software development projects. However, it is very difficult for the software managers to select the optimal software reliability model for the actual software development project. As an example, the software managers can assess the software reliability for the past data sets by using usual model evaluation criteria such as the Akaike's information criterion, mean square errors, predicted relative errors, and so on. On the other hand, the estimation results based on the past fault data cannot guarantee for the prediction of future data sets in actual software projects.

This paper focuses on the learning for the software development projects. We have proposed the selection method of optimal software reliability model based on the deep learning. In particular, it is difficult to assess the reliability by only using the past fault data, because the estimation results based on the past fault data cannot be guaranteed for the future data sets of actual software projects. In this paper, we have compared the methods of reliability assessment based on neural network with that of deep learning. Moreover, we have shown that the proposed method based on the deep learning can assess better than that based on neural network. Thereby, we have found that our method can assess the software reliability in the future with high accuracy based on the past software project data.

Furthermore, we have discussed the optimal release time and total expected software cost in terms of the model selection based on the deep learning. In particular, we have found that it is very important to decide the best model in terms of optimal release time, because the model selection has a great influence on the optimal release time and total software cost.

Acknowledgments

This work was supported in part by the Telecommunications Advancement Foundation in Japan, the Okawa Foundation for Information and Telecommunications in Japan, and the JSPS KAKENHI Grant No. 15K00102 and No. 16K01242 in Japan.

References

[1] Lyu, M. R., ed. (1996). *Handbook of Software Reliability Engineering.* Los Alamitos, CA: IEEE Computer Society Press.

[2] Yamada, S. (2014). *Software Reliability Modeling: Fundamentals and Applications.* Tokyo/Heidelberg: Springer.

[3] Kapur, P. K., Pham, H., Gupta, A., and Jha, P. C. (2011). *Software Reliability Assessment with or Applications.* London: Springer.

[4] Yamada, S., and Osaki, S. (1985). Cost-reliability optimal software release policies for software systems. *IEEE Trans. Reliab.* R-34, 422–424.

[5] Yamada, S., and Osaki, S. (1987). Optimal software release policies with simultaneous cost and reliability requirements. *Eur J. Operat. Res.* 31, 46–51.

[6] Karnin, E. D. (1990). A simple procedure for pruning back-propagation trained neural networks. *IEEE Trans. Neural Netw.* 1, 239–242.

[7] Kingma, D. P., Rezende, D. J., Mohamed, S., Welling, M. (2014). "Semi-supervised learning with deep generative models," in *Proceedings of Neural Information Processing Systems*, 3581–3589.

[8] Blum, A., Lafferty, J., Rwebangira, M. R., and Reddy, R. (2004). "Semi-supervised learning using randomized mincuts," in *Proceedings of the International Conference on Machine Learning*, 1–13.

[9] George, E. D., Dong, Y., Li, D., and Alex, A. (2012). Context-dependent pre-trained deep neural networks for large-vocabulary speech recognition. *IEEE Trans. Audio Speech Lang. Process.* 20: 30–42.

[10] Vincent, P., Larochelle, H., Lajoie, I., Bengio, Y., and Manzagol, P. A. (2010). Stacked denoising autoencoders: learning useful representations in a deep network with a local denoising criterion. *J. Mach. Learn. Res.* 11, 3371–3408.

[11] Martinez, H. P., Bengio, Y., and Yannakakis, G. N. (2013). Learning deep physiological models of affect. *IEEE Comput. Intell. Mag.* 8, 20–33.

[12] Hutchinson, B., Deng, L., and Yu, D. (2013). Tensor deep stacking networks. *IEEE Trans Pattern Anal. Mach Intell.* 35, 1944–1957.

[13] Brooks, W. D., and Motley R. W. (1980). *Analysis of Discrete Software Reliability Models*. Technical Report RADC-TR-80-84, Rome Air Development Center, Berlin.

Biographies

Yoshinobu Tamura received the B.S.E., M.S., and Ph.D. degrees from Tottori University in 1998, 2000, and 2003, respectively. From 2003 to 2006, he was a Research Assistant at Tottori University of Environmental Studies. From 2006 to 2009, he was a Lecturer and Associate Professor at Faculty of Applied Information Science of Hiroshima Institute of Technology, Hiroshima, Japan. Since 2009, he has been working as an Associate Professor at Yamaguchi University, Ube, Japan. His research interests include reliability assessment for open source software. He is a regular member of the Institute of Electronics, Information and Communication Engineers of Japan, the Information Processing Society of Japan, the Operations Research Society of Japan, the Society of Project Management of Japan, and the IEEE. Dr. Tamura received the Presentation Award of the Seventh International Conference on Industrial Management in 2004, the IEEE Reliability Society Japan Chapter Awards in 2007, the Research Leadership Award in Area of Reliability from the ICRITO in 2010, and the Best Paper Award of the IEEE International Conference on Industrial Engineering and Engineering Management in 2012.

Shigeru Yamada received the B.S.E., M.S., and Ph.D. degrees from Hiroshima University, Japan, in 1975, 1977, and 1985, respectively. Since 1993, he has been working as a professor at the Department of Social Management Engineering, Graduate School of Engineering, Tottori University, Tottori-shi, Japan. He has published over 500 reviewed technical papers in the area of software reliability engineering, project management, reliability engineering, and quality control. He has authored several books entitled such as Introduction to Software Management Model (Kyoritsu Shuppan, 1993), Software Reliability Models: Fundamentals and Applications (JUSE, Tokyo, 1994), Statistical Quality Control for TQM (Corona Publishing, Tokyo, 1998), Software Reliability: Model, Tool, Management (The Society of Project Management, 2004), Quality-Oriented Software Management (Morikita Shuppan, 2007), Elements of Software Reliability Modeling Approach (Kyoritsu Shuppan, 2011), Project Management (Kyoritsu Shuppan, 2012), Software Engineering: Fundamentals and Applications (Suurikougaku Publishing, 2013), Software Reliability Modeling: Fundamentals and Applications (Springer-Verlag, 2014), and OSS Reliability Measurement and Assessment (Springer-Verlag, 2016). Dr. Yamada received the Best Author Award from the Information Processing Society of Japan in 1992, the

TELECOM System Technology Award from the Telecommunications Advancement Foundation in 1993, the Best Paper Award from the Reliability Engineering Association of Japan in 1999, the International Leadership Award in Reliability Engg. Research from the ICQRIT/SREQOM in 2003, the Best Paper Award at the 2004 International Computer Symposium, the Best Paper Award from the Society of Project Management in 2006, the Leadership Award from the ISSAT in 2007, the Outstanding Paper Award at the IEEE International Conference on Industrial Engineering and Engineering Management (IEEM208) in 2008, the International Leadership and Pioneering Research Award in Software Reliability Engineering from the SREQOM/ICQRIT in 2009, the Exceptional International Leadership and Contribution Award in Software Reliability at the ICRITO'2010, 2011 Best Paper Award from the IEEE Reliability Society Japan Chapter in 2012, the Leadership Award from the ISSAT in 2014, and the Project Management Service Award from the SPM in 2014. He is a regular member of the IEICE, the Information Processing Society of Japan, the Operations Research Society of Japan, the Reliability Engineering Association of Japan, Japan Industrial Management Association, the Japanese Society for Quality Control, the Society of Project Management, and the IEEE.

Location Routing Inventory Problem with Transhipment Points Using *p*-center

S. S. R. Shariff*, N. S. Kamal, M. Omar and N. H. Moin

Centre for Statistical and Decision Science Studies, Faculty of Computer and Mathematical Sciences, Universiti Teknologi MARA, 40450 Shah Alam, Selangor, Malaysia
*Corresponding Author: shari990@salam.uitm.edu.my

Received 4 September 2016; Accepted 18 October 2016;
Publication 29 October 2016

Abstract

Location Routing Inventory Problem with Transhipment (LRIP-T) is a collaboration of the three components in the supply chain which are location-allocation; vehicle routing and inventory management problems that allow transhipment process, in a way that the total system cost and the total operational time are minimized. This study is to determine a set of customer points to act as the transhipment point as and when it is necessary, based on the surplus quantities it has, the quantities to ship to the needing customers and the sequence in which customers are replenished by homogeneous fleet of vehicles. The transhipment point is selected from the existing customers using *p*-center. The performance of the selection is evaluated using a set of benchmark data and a real life data. For the real life data, sensitivity analysis based on the number of distribution centres and size of lorries are presented. Results show important savings achieved when compared to the existing model in solving the supply chain problem.

Keywords: Location Inventory Routing Problem with Transhipment, *p*-center.

Journal of Industrial Engineering and Management Science, Vol. 1, 59–72.
doi: 10.13052/jiems2446-1822.2016.004

1 Introduction

Location routing inventory problem (LRIP) is an integration of three keys logistics decision problems which are location-allocation, vehicle routing and inventory management. LRIP arises when decisions on the three problems must be taken simultaneously. According to [1, 3], LRIP in the distribution systems is to allocate depots from several potential locations to schedule the routes for vehicle in meeting the customer's demand and to determine the inventory policy based on the information of customer's demands in order to minimize the system's total costs. Location inventory routing problem (LRIP) is a branch of logistic study that is not widely explored by the researchers due to the interrelated decision areas. However, excellent integration consequently presents significant saving costs as simultaneous decisions imposed in solving the problems.

Not having enough stock to fulfill the customer's demand or stock out is one of issues in the supply chain distribution. This unexpected situation occurred when there is an excess demand that lead to falling of inventories. When this happened, customers tend to purchase goods at another store or does not purchase at all. In addition, when a substitution is made, the retailer and supplier lose their potential sale as customers start to switch to another substitute and they may experience customer's dissatisfaction, worst outcome is losing a customer. It is vice versa of surplus where surplus is described as a situation to retain excess inventory. Surplus can cause profit loss to a company. When there is a surplus, it takes up space and increased holding costs. We have seen the impact of both issues in supply chain, thus we consider incorporating the transshipment process in order to prevent it from happening.

Transshipment has important role in the supply chain as it allows the goods to be shipped from depot to a customer or from a customer with excess stock to be shipped to another customer. It is profitable to the supplier as they can save the transportation cost in terms of delivering goods to the respective customer and at the same time, customer can also save cost in terms of storage space and obsolete the access stock. Several studies implemented transshipment and proved that it is effective compared with the LRIP without transshipment. Researchers tend to integrate transshipment with inventory routing problem (IRP) such as researchers [2, 4–6]. IRP is the combination of two problems in operation research; vehicle routing and inventory and IRPT is the extension of IRP with insertion of transshipment. In both studies they used the customers as the transshipment points and we also implemented the similar technique. The transshipment is only allowed when the location of transshipment point with retailer are nearer compared to the location of supplier. It is profitable to the supplier as they can save the transportation cost because of the short distance of travel to the customers. In this study, we consider LRIP-T or Location Routing Inventory Problem with Transshipment by applying p-center to select the transshipment center and apply the model to solve a real world problem.

2 Research Methods

Model

The model used is adapted from the earlier study [9] in which the transshipment cost is incorporated into total system cost of existing model adapted from [8]. p-center model is used to select the best transshipment point. In p-center problem the maximum distance between the distribution point and customer's node is minimized.

where

r_{ij}: the cost (unit) of transshipping product from i to j

Q_{vij}: the quantity transshipped from customer i in period t

MaxD: the total cost at the maximum distance,

$$\text{Maximize} \sum_{t=1}^{l} \sum_{k=1}^{K} \sum_{h=1}^{N+M} \sum_{g=1}^{N+M} D_{hg} \cdot W_{hgkt}$$

The formulation of the objective function is as below:

Minimize

$$MaxD + \sum_{j=N+1}^{N+M} \left(E_j + \left(\sum_{t=1}^{l} (B_j + Y_j) H_j(t) + q_j \cdot \frac{P}{l} \cdot \right. \right.$$

$$\left. \left. \sum_{i=1}^{l} (V_j(t) + \alpha_j \gamma_j(t)) + B_j \right) \cdot A_j \right.$$

Subject to:

$$\sum_{k=1}^{K} \sum_{h=1}^{N+M} W_{gikt} = 1, \text{for all } i, t \tag{1}$$

$$\sum_{g=1}^{N+M} W_{ghki} - \sum_{g=1}^{N+M} W_{hgki} = 0, \text{for all } h, k, \text{and } I \tag{2}$$

$$\sum_{h=1}^{N+M} \sum_{g=1}^{N+M} W_{ghki} \leq 1, \text{for all } k, \text{and } t \tag{3}$$

$$\beta_k^t(i) \leq CK, \text{where } i \in \{0, 1, 2, \ldots, N\} \tag{4}$$

$$\tau_i(t) \leq U(f_i(t)), \text{ for all } I \tag{5}$$

$$\sum_{h=1}^{N+M} W_{ihkt} - \sum_{h=1}^{N+M} W_{jhkt} - S_{ij} \leq 1, \text{ for all } i, j, k \text{ and } t \tag{6}$$

$$W_{ghkt} \in \{0, 1\}, \text{for all } g, h, k \text{ and } t \tag{7}$$

$$A_j \in \{0, 1\}, \text{ for all } j \tag{8}$$

$$S_{ij} \in \{0, 1\}, \text{ for all } i \text{ and } j \tag{9}$$

$$O_{jkt} \in \{0, 1\}, \text{ for all } j, k, \text{ and } t \tag{10}$$

$$O_j \in \{0, 1\}, \text{ for all } j \tag{11}$$

The objective function is to minimize the total system cost expressed by the summation of logistics center's set up cost, transportation cost and inventory cost. Constraint (1) ensures that each customer appears in only one route during period t. Constraint (2) states that every point entered should be the same point the vehicle leaves. Constraint (3) insures that each route only served by one logistics center. Constraint (4) and (5) are the statement for any period, in any point the total load is less than the vehicle capacity and the actual volume is equal or less than expected volume. Constraint (6) states that a customer can be allocated to a logistics center if there is a route passed by the customer. Constraint (7–11) insures the decision variable's integrality.

The following notations are used to describe the formulation.

N: number of customer points
M: number of logistics centers
h: index of customer point or logistic centers $(1 \leq h \leq N + M)$
g: index of customer point or logistic centers $(1 \leq g \leq N + M)$
i: index of customer point $(1 \leq i \leq N)$

j: index of logistics center $(N + 1 \leq j \leq N + M)$
k: index of vehicles or routes $(1 \leq k \leq K)$
t: index of time periods of planning horizon (p) $(1 \leq t \leq p)$
E_j: cost to establish the logistics center j
D_{hg}: distance between point h and point g
P: length of the planning horizon
CK: capacity of vehicles
c: unit cost of vehicles
B_j: cost of dispatching the product from factory to logistics center j
α_j: probability of being reused after a circulation
q_j: holding cost of products (in unit) at logistics center j
Y_j: ordering cost of every time in logistics center j
L_j: lead time in logistics center j where, $L_j \leq P/p$
$V_j(t)$: logistics center's j inventory level start at period t, where $V_j(t = 0) = 0$ stands for inventory is zero at the initial stage of planning horizon

And the decision variables are:

W_{hgkt}: 1, if point h immediately go to point g on route k in period t; 0, otherwise

S_{ij}: 1, if customer i is allocated to logistics center j; 0, otherwise

A_j: 1, if logistics center j is opened; 0, otherwise

O_{jkt}: 1, if route k is served by logistics center j in period t; 0, otherwise

$H_j(t)$: 1, if there exists an order for new product at logistics center j; 0, otherwise

$R_i(t)$: actual collection volume from customer i in period t

3 Data Analysis & Results

The analysis is divided into two sections: (1) on benchmark data for method validation and (2) on real life data for application.

3.1 Data Preparation (Benchmark Data)

Data are adapted based on [8, 9] and used based on the following assumptions. First, there are 5 candidates of logistics centers and 15 customers, the capacity of vehicles (CK) is 125 units, the unit cost of vehicles (c) is 1/unit distance and the probability of being reused after a circulation (α_j) is 0.9. In addition, we assume the expected distribution demand is randomly populated and obey the Poisson distribution with $\lambda_i = 2$ and $\delta_i = 1$. Tables 1 and 2 present the parameter values of logistics centers and customer points.

Table 1   Logistics centers

Logistics Centers	Coordinates	α_j	Setup Cost
D1	(43,49)	25	10,000
D2	(1,12)	20	12,000
D3	(41,30)	30	14,000
D4	(5,58)	10	13,000
D5	(24,19)	15	12,000

Table 2   Customer points

Customer Points	Coordinates	Demand
C1	(15,3)	42
C2	(18,24)	56
C3	(2,59)	53
C4	(9,6)	34
C5	(49,54)	48
C6	(33,10)	61
C7	(30,50)	37
C8	(24,59)	45
C9	(3,35)	48
C10	(33,21)	68
C11	(45,27)	32
C12	(46,6)	69
C13	(24,32)	45
C14	(28,33)	62
C15	(2,0)	47
Total		747

3.2 Results on Benchmark Data

In earlier study, the transshipment point is determined based on *p*-center. In the actual scenario, the total system cost is the summation of establishing cost of logistics centers, transportation cost and inventory cost. There are some modified costs for LRIP-T in this study.

  i. Logistics Center Establishment Cost – No establishment cost needed as the transshipment center will be selected among the customers.
 ii. Transportation cost – Transportation cost covers the from the distribution centers and from the transshipment centers.
iii. Inventory cost – The inventory level at the transshipment center is assumed to be zero, or the center is assumed not to keep any stock.

Table 3 shows that the logistics center D1 is determined to be the best center with the lowest average travelled distance of 96.67 km when the supply is

Table 3 Results for limited supply

Logistics Center	Total Distance (km)	Average Distance (km)	Average Distance/Vehicle (km)	Average Waste (unit)
D1	690.69	46.05	230.23	51
D2	725.91	48.39	241.97	51
D3	1022.31	68.15	340.77	51
D4	956.7	63.78	318.9	51
D5	749.79	49.99	249.93	51

unlimited and 46.05 km only when the supply is limited [9]. Considering the average traveled distance at 46.05 km, this means that the total cost is equal to 46.05 x 747 = 34399.35.

In the second phase, a transshipment point is determined based on the notion of *p*-center. The transshipment process is considered only for the case when supply is limited.

Initial sets of routes are chosen and MaxD is calculated and minimized. Each customer from C1 up till C15 has equal chance of being chosen as the transshipment point. Total transportation system cost is calculated and the best result is based on the set that produces the lowest cost.

Table 4 shows the final answer that lists the customers that being covered by D1 as the logistics center and those that being covered by C11 as the transshipment point. This also shows the total delivery cost (in km) of 19, 756.8 compared to 34399.35. This means that LRIP-T model has performed tremendously well in order to save the total costs for all the three systems; location, inventory and routing.

Table 4 Coverage of logistics center and the transshipment point

D1 Coverage		C11 Coverage	
Customer Points	Cost	Customer Points	Cost
C2	2287.6	C1	775.6
C3	2276.4	C5	336.0
C4	1867.6	C9	700.0
C6	1554.0	C14	400.4
C7	2632.0	C15	532.0
C8	1920.8		
C10	1103.2		
C12	2105.6		
C13	1265.6		
	Total cost = 19756.8		

3.3 Real-Life Data

In order to apply the model, a case study on delivery of frozen chicken by a local company in Selangor, Malaysia is chosen. The main distribution centre is located in Semenyih, Selangor, however, the model is applied to solve its delivery problem to its 36 customers in another state, Pahang, Malaysia. Figure 1 shows a map of chicken order for a month in Pahang, Malaysia and the customers are clustered in seven (7) areas, as some (for example C6–C28) customers are quite close to each other. There are a lot of independent agents who operate from grocery shops or homes in each

Figure 1   Map of customers location.

area which lead to the unavailability of the data. Therefore, data cleaning is done by combining all of the customers in different locations to be within the same clustered area. For the present case study, the analyses are divided into two phases. The first phase is having all chickens delivered from the main distribution centre and described as One Depot. In the analysis the best planning route for delivery is chosen while minimizing the total travel distance and at the same time determining the number of lorry use. The second phase is when the transhipment point is identified using p-centre and to be the second distribution centre and described as Two Depot. Similarly, the best planning route for delivery is chosen while minimizing the total travel distance and at the same time determining the number of lorry use.

Table 5 shows the vehicle type which is the size of lorry such as small lorry and large lorry, capacity of the lorry and cost per delivery. Each customer is served by one or more facilities with specified transportation links. Moreover, each outsource transportation has a fixed capacity to deliver customers demand. Furthermore, the total chicken out per delivery takes about maximum 1000 of chickens when customers request for the small lorry and maximum 2300 of chickens when customers request for the large lorry, based on customers' demands. In addition, each delivery is charged about RM 500 per delivery for small lorry otherwise each delivery is charged about RM 800 per delivery for large lorry.

3.3.1 Result for Phase 1 – One Depot
Tables 6 and 7 show the results when all chickens are delivered from the main distribution centre or one depot only.

3.3.2 Result for Phase 2 – Two Depots
The transhipment point C1 is chosen using p-centre and the chickens are delivered from both the main distribution centre (0) and from C1. The best route is determined using Excel Solver and 15 customers get their delivery directly from the main distributor while the balance two (12) customers get it from the transhipment point C1. Note that C1 also gets its supply from the main distributor. Total distance travel by all chickens are summarised in Tables 8 and 9.

Table 5 Vehicle type, capacity of lorry and cost per delivery

Vehicle Type (Size of Lorry)	Capacity of the Lorry	Fixed Cost Per Delivery
Small	1000 chickens per delivery	RM 500
Large	2300 chickens per delivery	RM 800

Table 6   Average traveled distance using small lorry

Number of Lorry	Routes	No. of Chickens	Distance Traveled
First	Depot-C13	1000	326700
Second	Depot-C13	1000	326700
Third	Depot-C6-C7-C10-C11-C12-C14-C17-C18-C19-C21-C23	740	241758
Forth	Depot-C13-C30-C32-C1	850	326647
Fifth	Depot-C29-C36	130	19686
Total		3720	1241491
Average travel distance per chicken			333.73

Table 7   Average traveled distance using large lorry

Number of Lorry	Routes	No. of Chickens Per Delivery	
First	Depot-C13	2300	751410
Second	Depot-C6-C7-C10-C11-C12-C13-C14-C17-C18-C19-C21-C23-C30-C32-C1-C29-C36	1420	484049
Total		3720	1235459
Average travel distance per chicken			332.11

Table 8   Delivery from transhipment point C1

No. of Lorry	Routes	No. of Chickens	Total Distance
First	Depot-C1-C13	2640	121440
Second	Depot-C1-C6-C7-C10-C11-C12-C13-C14-C17-C18-C19-C21-C23-C30-C32	950	165550
Total		3590	286990
Average travel distance by each chicken			79.94

Table 9   Delivery from main depot

Number of Lorry	Routes	No. of Chickens	Total Distance
First	Depot-C36-C29	130	4316
Total		130	4316
Average travel distance by each chicken			33.2

In Section 3.3.1, the results show that the size of lorry does not contribute to lower distance traveled by each unit of the chicken. However, comparing the results in Sections 3.3.1 and 3.3.2, we can see that by having the transhipment point the average traveled distance can be reduced tremendously from 333.73 (using small lorry) and 332.11 (using large lorry) to only 79.94 from the transhipment point and only 33.2 from the main depot.

4 Conclusion

In this paper, we have proposed a new way of determining the transshipment point among the customers for location routing inventory problem. The selection of routes in this analysis are randomly picked and tested in order to achieve the maximum distance that has the minimum total system cost. The total system cost has considered the new technique of choosing the transshipment point using p-center. This technique has excluded some costs and only has the total distance travelled in the formulation and applied into the real life problem and see tremendous improvement. In future research, we will further consider using the heuristic method called genetic algorithm for a better route's selection. Another area of possible research involves LRIP-T together with development of new algorithm is the stochastic demand. For real life problem, fluctuation of customer's demand also needs to be considered.

Acknowledgement

We would like to acknowledge Ministry of Higher Education Malaysia for funding this research through the Research Management Institute (RMI) of Universiti Teknologi MARA, Malaysia, Grant No: 600-RMI/FRGS 5/3(9/2013)) and ICatse for the invitation of this special issue.

References

[1] Xuefeng, W. (2010). "An integrated multi-depot location-inventory-routing problem for logistics distribution system planning of a chain enterprise" in *Logistics Systems and Intelligent Management, 2010 International Conference on 3*, Nanchang China: IEEE, 1427–1431.

[2] Granada, M. G., and Silva, C. W. (2012). "Inventory location routing problem: a column generation approach." in *Proceedings of the 2012 International Conference on Industrial Engineering and Operations Management,* Istanbul, Turkey, 482–491.

[3] Bertazzi, L., Paletta, G., and Speranza, M. G. (2002). *Transportation Science,* 36, 119–132.

[4] Kleywegt, A. J., Nori, V. S., and Savelsbergh, M. W. P. (2002). *Transportation Science,* 36, 94–118.

[5] Zhao, Q. H. (2003). "Study on logistics optimization models", PhD thesis, Beihang University.

[6] Yang, F. M., and Xiao, H. J. (2007). *Systems Engineering, Theory & Practice*, 27, 28–35.
[7] Zhang, B., Ma, Z., and Jiang, S. (2008). "Location routing inventory problem with stochastic demand in logistics distribution systems", In *Proceeding of: Wireless Communications, Networking and Mobile Computing*, IEEE Xplore, 1–4.
[8] Ahmad, H., Hamzah, P., Md Yasin, Z. A. M., and Shariff, S. S. R. (2014). "Location Routing Inventory Problem with Transshipment (LRIP-T)" in *Proceedings of the 2014 International Conference on Industrial Engineering and Operations Management*, Bali, Indonesia, 1595–1605.
[9] Shariff, S. S. R., Omar, M., and Moin, N. H. (2016). "Location Routing Inventory Problem with Transshipment Points Using p-Center", in *2016 International Conference on Industrial Engineering, Management Science and Application (ICIMSA)*, Jeju, South Korea, 1–5.

Biographies

S. S. R. Shariff is currently a fulltime senior lecturer at the Centre for Statistical and Decision Science Studies at the Faculty of Computer and Mathematical Sciences of Universiti Teknologi MARA (UiTM), Shah Alam, Malaysia. Dr. Shariff earned a Bachelor of Science degree in Statistics and Mathematics from Purdue University, West Lafayette, Indiana, USA, a Master in Information Technology from UiTM, Shah Alam and Ph.D. in Operational Research from University of Malaya, Kuala Lumpur. She is the Head of Research Interest Group for Logistics Modelling and has published and reviewed journal and conference papers. Her research interests include facility location modeling, logistics process modeling, warehousing, inventory routing, optimization and performance measurement. She is the Secretary for Management Science/Operations Research Society of Malaysia (2013–2017), a member of INFORMS, and a Lifetime Member of Mathematical Society of Malaysia (PERSAMA).

N. S. Kamal is a final year student at UiTM, pursuing a Master degree in Quantitative Sciences. Upon graduation, she works as a planner in a company.

M. Omar is a full Professor at Institute of Mathematical Sciences, University of Malaya, Kuala Lumpur. His research interests include but not limited to Industrial Optimisation and Control (Production Planning, Vendor-Buyer, Optimal Policy.

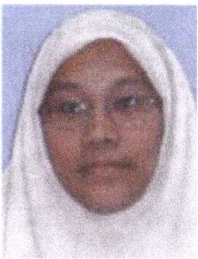

N. H. Moin is an Associate Professor at Institute of Mathematical Sciences University of Malaya, Kuala Lumpur. Her research interests are Vehicle

Routing, Inventory Routing, Production Inventory Distribution Routing Problem, Scheduling, Metaheuristics (Genetic Algorithm, Ant Colony, Artificial Bee, Variable Neighborhood Search), Matheuristic (Combination of Linear Programming Technique and Metaheuristic). (Vehicle Routing, Inventory Routing, Production Inventory Distribution Routing Problem, Scheduling, Metaheuristics, Matheuristic)

Design of Arc Fault Detection Circuit in Low Voltage Switchboard

Kuan Lee Choo and Ahmad Azri Sa'adon

Faculty of Engineering and Technology Infrastructure, Infrastructure University Kuala Lumpur, Selangor, Malaysia
E-mail: lckuan@iukl.edu.my; azriaces@gmail.com

Received 5 October 2016; Accepted 17 November 2016;
Publication 26 November 2016

Abstract

This paper presents the design of the arc fault detection circuit that is able to detect three different signals prior to the occurrence of an arcing fault in the low voltage switchboard. There are pressure, heat and light signals. The simulation results show that the proposed arc fault detection circuit will activate the relay and send a trip signal to the circuit breaker if the illumination level in the interior of the low voltage (LV) switchboard is more than a predetermined value and at the same time both the pressure and temperature detectors detect a pressure and temperature level which are higher than the reference value. This is to ensure that no fault tripping signal is sent to the circuit breaker and therefore avoid unnecessary power shut down.

Keywords: LM 335 Temperature Sensor, 1140 Pressure Sensor, ISL2910 Light to Analog Sensor, Arc Fault, Low Voltage Switchboard.

1 Introduction

An arc fault is a high power discharge of electricity between two or more conductors. This discharge translates into heat, which can break down the wire's insulation and possibly trigger an electrical fire. These arc faults can range in power from a few amps up to thousands of amps high and are highly

Journal of Industrial Engineering and Management Science, Vol. 1, 73–88.
doi: 10.13052/jiems2446-1822.2016.005

variable in terms of strength and duration. Common causes of arc faults include faulty connections due to corrosion and faulty initial installation. The number of incidents related to arcing fault began to rise in the 1960's when the field power system undergoes rapid development to meet the increasing of load demand [1].

Arc incidents occur due to various reasons such as, poorly installed equipment (human mistakes), natural aging of equipment, bad connections, faulty connection due to corrosion. Statistics have shown that 80% of electrically related accidents and fatalities involving qualified workers are caused by arc flash or arc blast. A true arc fault will rapidly increase energy level up to 20MW cycle, increase in pressure up to 3 atm, and in heat up to 3000 degrees Celsius [2].

An arcing fault instantaneously release large amount of radiant energy. The radiant energy includes both light and thermal energy. Light intensities can be thousands of time higher than normal ambient light [3]. The light sensor is incorporated in the arc fault detection circuit as light is relatively easy to detect [4]. The main disadvantage of a light detection circuit is the risk of tripping from a light source not related to an arc flash. However, this risk can be reduced to minimum in this proposed design that incorporates three different signals prior to the occurrence of an arcing fault in the low voltage switchboard. Therefore, it is able to ensure no fault tripping signal is sent to the circuit breaker and detect the possibility of arc fault occurrence at the early stage.

This paper is organized as follows: Section 2 provides the study of behavior and characteristic of the arc fault. Section 3 subsequently presents the system description of the arc fault detection circuit. Section 4 describes the modeling of the arc fault detector circuit. Section 5 provides the simulation results of the arc fault detection circuit. Lastly, Section 6 concludes the finding of this paper.

2 Behaviour and Characteristic of Arc Fault

An arc fault is the discharge of electricity through the air between two conductors which creates huge quantities of heat and light. It is a high resistance fault with resistance similar to many loads and it is a time varying resistor which can dissipate large amount of heat in the switchboard [5].

Circuit breakers are tested by bolting a heavy metallic short across the output terminals to determine their capabilities of handling an essentially zero resistance load. The zero resistance faults are named as bolted fault.

Bolted fault current is the highest possible current supplied by the source and a protective system is designed according to the value of bolted fault current. The protective system must be able to detect the bolted fault and the protective devices must be capable of interrupting this value of current [6].

Due to the high resistance loads, an arcing fault will result in much lower values of current. The arcing fault current is often insufficient to operate the protective devices such as circuit breakers, fuses and relays. As a result, the arcing fault will persist until severe burn down damage occurs. The magnitude of the arc current is limited by the resistance of the arc and the impedance of the ground path [7].

Arc faults are categorized into series arc faults and parallel arc faults. Series arc faults happen when the current carrying paths in series with the loads are unintentionally broken whereas parallel arc faults happen between two phases, phase to ground or phase to neutral of the switchboard [8].

Large amounts of heat will be dissipated during an arc event. A portion of this heat is coupled directly into the conductors, a portion heats the air and another portion is radiated in various optical wavelengths. Hasty heating of the air and the expansion of the vaporized metal into gas produces a strong pressure wave which will blow off the covers of the switchboards and collapse the substations [8].

Figure 1 shows the time, current and damage for the 53 arcing tests. When the circuit breakers are tripped within less than 0.25 seconds, the damage will be limited to smoke damage. The triangle markers represent arcs that cause only smoke damage to the side of switchboards. The square markers represent arcs that cause surface damage to the side of switchboards whereas the star pointers represent holes of several square inches at the side of the switchboards [5].

When an arc is ignited, the plasma cloud expands cylindrically around the arc. The expansion of the plasma is constrained by the parallel busbar and thus the plasma expands more to the front and the back of the bus. As the plasma reaches any obstructions such as the switchboard, plasma expansion is retarded by the obstructions. Due to the lower velocity of the arc, the plasma becomes more concentrated and its temperature and current will increase [5].

The root of the arc where the arc contacts the conductor is reported to reach temperatures exceeding 20000°C, whereas the plasma portion or positive column of the arc is around 13000°C [9]. For reference, surface of the sun is reported to be about 5000°C. The components in the switchboard can only withstand this temperature within 250 milliseconds before sustaining severe damages [10].

Damage vs. Time and Current

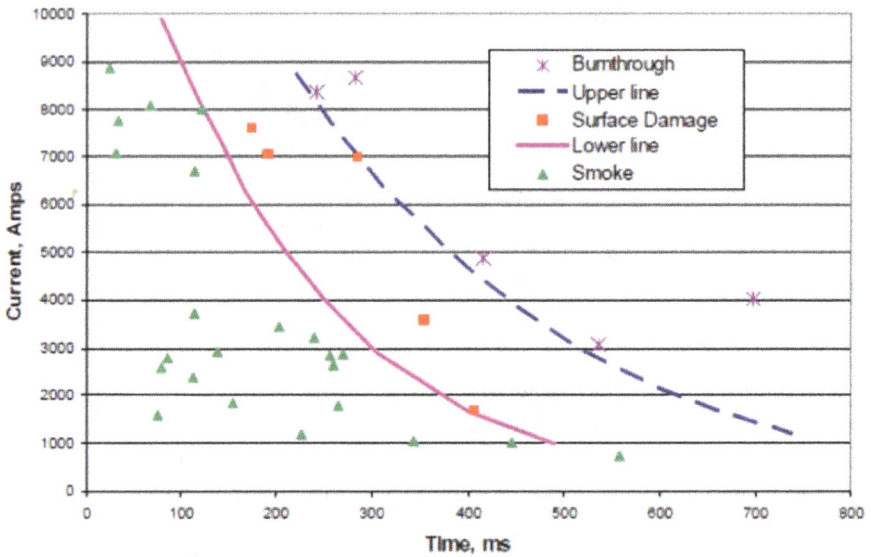

Figure 1   Damage to the side of a switchboard versus arc current and time.

The design of light detector is based on the principle that during an arc flash, large amount of light will be detected by the detector to indicate a flash. An arc flash can draw a fraction of bolted-fault current, especially in the early stages, circuit breakers alone cannot be relied upon to differentiate between the arcing current and a typical short-circuit current. By incorporating a light detector in the arc fault detection circuit, it reduces the total clearing time and the amount of energy released through an arcing fault greatly [11].

3 System Description of Arc Fault Detection Circuit

Figure 2 shows the block diagram for the proposed design of the arc fault detection circuit. It consists of a pressure detector, a temperature detector, a light detector, voltage comparator, an AND gate, a relay and a LED. The pressure, temperature and light detectors will detect the arc fault by amount of pressure heat and the light produced and convert into corresponding voltage and the output of voltage comparator from these sensors will send to the AND gate. A 'HIGH' output from the AND gate will trigger the relay to send a trip signal that will turn on the LED.

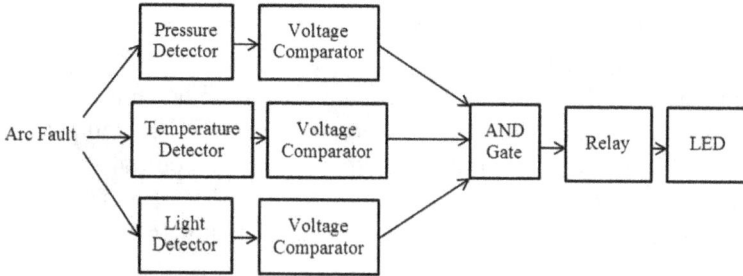

Figure 2   Block diagram of an arc fault detection circuit.

The 1140 pressure sensor is used to detect the presence of an arcing fault by sensing the pressure changes in the switchboard. 1140 has a breakdown voltage directly proportional to the temperature, which is +12 mV/kPa. 1140 is chosen because the 1140 Absolute Pressure Sensor is an air pressure sensor that measures the absolute pressure of its environment. It can measure pressures from 20 kPa to 400 kPa. In addition, it has a high precision and narrow measurement range compared to other types of pressure sensors. It has typically less than ±1.5% error. The pressure range for 1140 is 20 kPa to 400 kPa. In other words, the output voltage of this pressure sensor will range from 240 mV to 4.8 V.

The calculations for the output voltage are shown below:

$$\text{Output voltage for 20 kPa} = 20 \times 12 \times 10^{-3} = 0.24\,\text{V} \quad (1)$$

$$\text{Output voltage for 400 kPa} = 400 \times 12 \times 10^{-3} = 4.8\,\text{V} \quad (2)$$

Every 1 kPa increases in the pressure of the surrounding. The output voltage will increase by 12 mV. Since the atmospheric pressure is 100 kPa, it is assumed that under normal condition, the pressure inside a switchboard is about 150 kPa which is equivalent to 1.8 V. From [1], a true arc fault will rapidly increase in pressure up to 3 atm or equivalently to 303.975 kPa. Therefore, in this proposed design, the pressure of 250 kPa is set as the reference value. By calculation, the voltage value corresponding to 250 kPa is 3.0 V. This voltage is set as the reference voltage. The input voltage (i.e. the pressure inside the switchboard) will be compared with the reference voltage (i.e. the reference pressure value of 250 kPa). A "HIGH" output from the voltage comparator will send to the AND gate if the input voltage is higher than the 3.0 V.

The LM 335 temperature sensor is used to detect the presence of an arcing fault by sensing the temperature changes in the switchboard. LM 335 has a breakdown voltage directly proportional to the temperature, which is +10 mV/°K. LM 335 is chosen because it is precise, easily calibrated and integrated circuit temperature sensor. In addition, it has a linear output and it is cheaper compared to other types of temperature sensors. When it is calibrated at 25°C, it has typically less than 1°C error over a 100°C. The temperature range for LM 335 is –40°C to 100°C. In other words, the output voltage of this temperature sensor will range from 2.33 V to 3.73 V.

The calculations for the output voltage are shown below [6]:

$$\text{Output voltage for } -40°C = (-40 + 273) \times 10 \times 10^{-3}$$
$$= 2.33 \text{ V} \qquad (3)$$

$$\text{Output voltage for } 100°C = (100 + 273) \times 10 \times 10^{-3} = 3.73 \text{ V} \quad (4)$$

For every 1°K increase, the output voltage will increase by 10 mV. Under normal condition, the temperature inside a switchboard is about 40°C. It is assumed that the temperature inside a switchboard will rise to 92°C and by calculation; the corresponding voltage is 3.65 V. This voltage is set as the reference voltage. The input voltage will be compared with the reference voltage. The relay will be activated, once the input voltage is more than the reference voltage.

The ISL29101 light to analog sensor is used to detect the presence of an arcing fault by sensing the illumination (lux) level in the interior of the switchboard. It is able to detect up to an illumination level of 10000 lux. The corresponding output voltage for 0 lux to 100 lux is 1 mV to 1.65 V. Therefore, for every 1lux increase, the corresponding voltage is 16.5 mV.

Since the illumination level for the switchroom from IES standard is 300 lux [12], it is assumed that under normal condition, the illumination level in the interior of the switchboard is about 50 lux which is equivalent to 0.825 V. In this proposed design, the illumination level of 90 lux is set as the reference value. And the voltage value corresponding to 90 lux is 1.5 V. This voltage is set as the reference voltage. The input voltage (i.e. the illumination level inside the switchboard) will be compared with the reference voltage (i.e. the reference illumination level of 90 lux). A "HIGH" output from the voltage comparator will send to the AND gate if the input voltage is higher than the 1.5 V.

4 Modelling of Arc Fault Detection Circuit

A buffer amplifier provides electrical impedance transformation from one circuit to another circuit. It is used to transfer a voltage from the light sensor, pressure sensor and temperature sensor to the voltage comparator. A unity gain buffer is used in the circuit design. The output of the op-amp (buffer) is connected to its inverting input, which is the negative feedback. Therefore, the output voltage is simply equal to the input voltage of the buffer. The outputs from the pressure sensor, temperature sensor and light sensor are connected to the non-inverting input of the buffer (op-amp), which is the positive feedback, and the output from the buffer is identical to the these sensors.

The pressure sensor, temperature sensor and light sensor will generate a voltage based signal with respect to the amount of pressure, temperature and light detected from the interior of the switchboard. The signal is then sent to a voltage comparator through a buffer. The voltage comparator is used to compare the signal with a reference voltage. The output of the comparator from all these sensors will produce a positive ('HIGH' output) value which will then become the inputs for the 7410 AND gate. Then the output of the AND gate will activate the relay and send a trip signal to the circuit breaker (i.e.to turn on the LED) if all the light sensor, pressure sensor and temperature sensor detect an illumination level, pressure and temperature that exceed the reference voltage of the voltage comparator. Else, the output voltage of the voltage comparator will indicate a negative value ('LOW' output) which will not trigger the trip signal.

Before detect the changes of light, pressure and temperature in the interior of switchboard and operate the LED when the predetermined limit is exceeded, these detectors are modeled to lower values of pressure and temperature with respect to the practical illumination level, pressure and temperature value.

The schematic diagram of the pressure detector, temperature detector and light detector in PSpice program are shown in Figures 3–5 respectively.

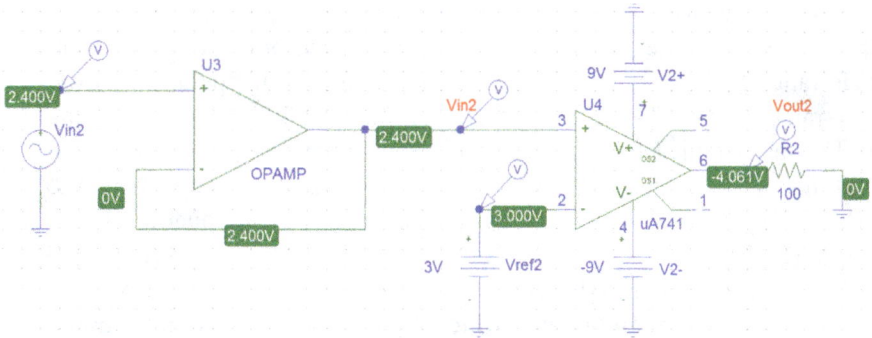

Figure 3 PSpice schematic diagram of the arc fault pressure detector.

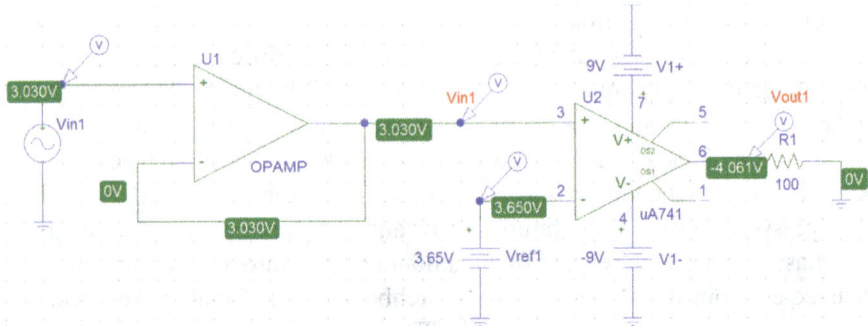

Figure 4   PSpice schematic diagram of the arc fault temperature detector.

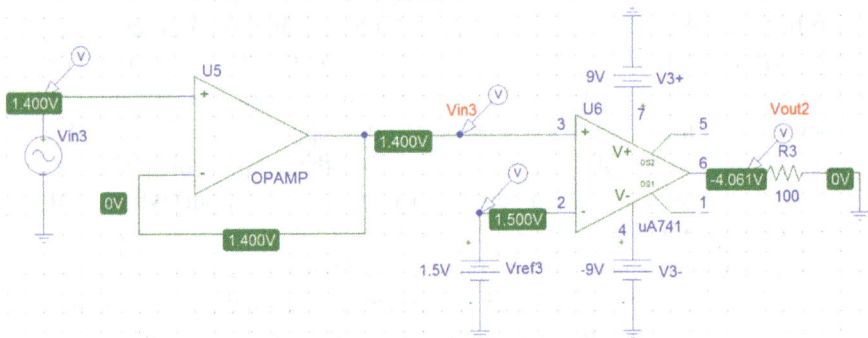

Figure 5   PSpice schematic diagram of the arc fault light detector.

Figure 6 shows the schematic diagram of the overall arc fault detection circuit using PSpice program. The input for this circuit is an AC supply. An AC supply is used to represent the output signal from the light, pressure and temperature sensor. The output voltage range of the sensor is used as the input voltage range for the circuit. The AC input voltage for temperature sensor, Vin1, is ranged from 2.33 V (corresponding to –40°C) to 3.73 V (corresponding to 100°C) as obtained from Equation (3) and Equation (4). The AC input voltage for pressure sensor, Vin2, is ranged from 0.2 mV (corresponding to 20 kPa) to 4.8 V (corresponding to 400 kPa) as obtained from Equation (1) and Equation (2). The AC input voltage for temperature sensor, Vin3, is ranged from 0 V (corresponding to 0 lux) to 1.65 V (corresponding to 100 lux).

U1, U3 and U5 are an op-amp, which represents a buffer in this circuit. The AC supply is connected to the positive feedback of U1, U3 and U5 and the negative feedback of U1 and U3 is connected to the output of U1, U3 and

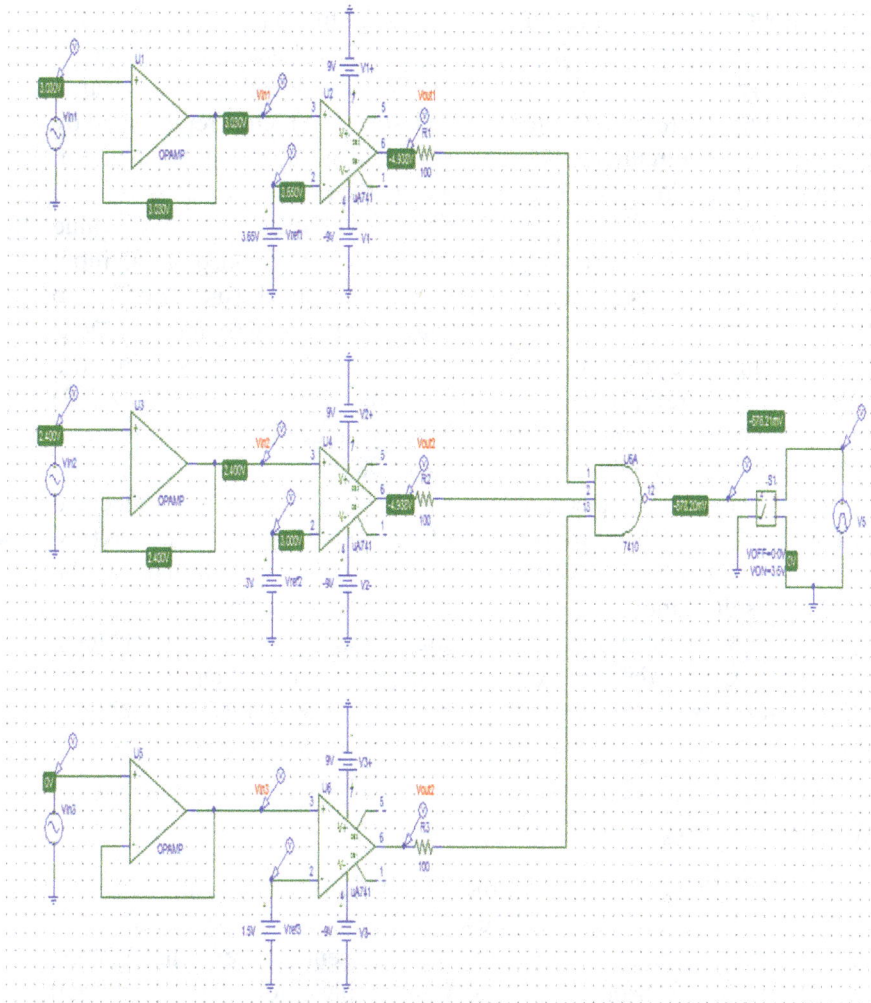

Figure 6 PSpice schematic diagram of the overall arc fault detection circuit.

U5 to produce a unity gain buffer. The output voltage of U1, U3 and U5 are same as the input voltage since it is a unity gain buffer.

Then, the output voltage of U1, Vin1, is connected to the positive feedback (pin 3) of U2, which is a voltage comparator. Also, the output voltage of U3, Vin2, is connected to the positive feedback (pin 3) of U4, which is a voltage comparator. The output voltage of U5, Vin3, is connected to the positive feedback (pin 3) of U6, which is a voltage comparator.

The output voltage of the buffer is used as the input voltage of the comparator. Theoretically, the input voltage of U2, U4 and U6 are identical to the output voltage of U1, U3 and U5 and are also identical to the output voltage of the light, pressure and temperature sensor, which is represented by an AC source in this circuit. uA741 op-amp is used as the voltage comparator. A DC input voltage, Vref1, of 3.65 V is placed at the negative feedback (pin 2) of U2 to produce a constant value of reference voltage. The voltage value of 3.65 V is equivalent to the temperature value of 92°C. Also, a DC input voltage, Vref2, of 3 V is placed at the negative feedback (pin 2) of U4 to produce a constant value of reference voltage. The voltage value of 3 V is equivalent to the pressure value of 250 kPa. For the light detector circuit, a DC input voltage, Vref3, of 1.5 V is placed at the negative feedback (pin 2) of U6 to produce a constant value of reference voltage. The voltage value of 1.6 V is equivalent to the illumination value of 90 lux.

A +9 V DC supply is connected to pin 7 and a –9 V DC supply is connected to pin 4 of U2, U4 and U6 to supply voltage for this component. Output voltage from U2, U4 and U6 (pin 6), Vout1, Vout2 and Vout3, is used to indicate the comparison result of the input voltage and the reference voltage.

The 7410 AND Gate with three inputs are connected to the outputs voltage of U2, U4 and U6 then AND Gate output is connected to the Voltage Controlled Switch which act as a relay that will activate the Voltage Pulse and therefore turn on the LED if output voltage of the AND gate is positive.

5 Simulation Results

The PSpice simulation results from the schematic diagram of the pressure, temperature and light detectors are shown in Figures 7–9 respectively.

The straight line in red color in Figure 9 represents the reference voltage of 1.5 V for the light sensor. The sine wave in purple color is the input signal that used to trigger the light sensor. The square wave is the output from voltage comparator which trigger as logic '1' ('HIGH') at 4.0 V if the input voltage of sine wave is higher than the reference voltage and becomes logic '0' ('LOW') at –4.0 V if the input voltage of sine wave is lower than the reference voltage.

Figure 10 shows the PSpice simulation result from the schematic diagram of the arc fault pressure and temperature detectors circuit. As can be seen in Figure 10, the input voltage of temperature sensor of 3.03 V is represented by the sine waveform (in pink color) and input voltage of pressure sensor of 2.4 V is represented by another sine waveform (in light blue color). The reference voltages are the two straight lines at 3.65 V for temperature sensor

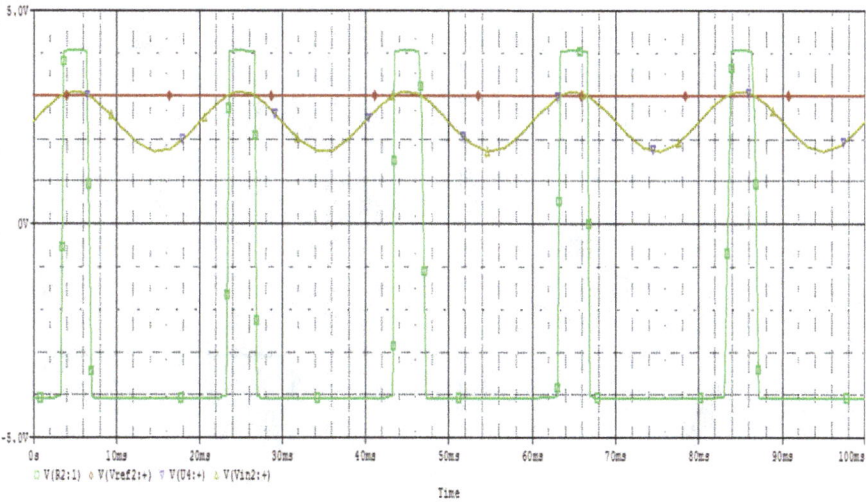

Figure 7 Simulation result of the arc fault pressure detector.

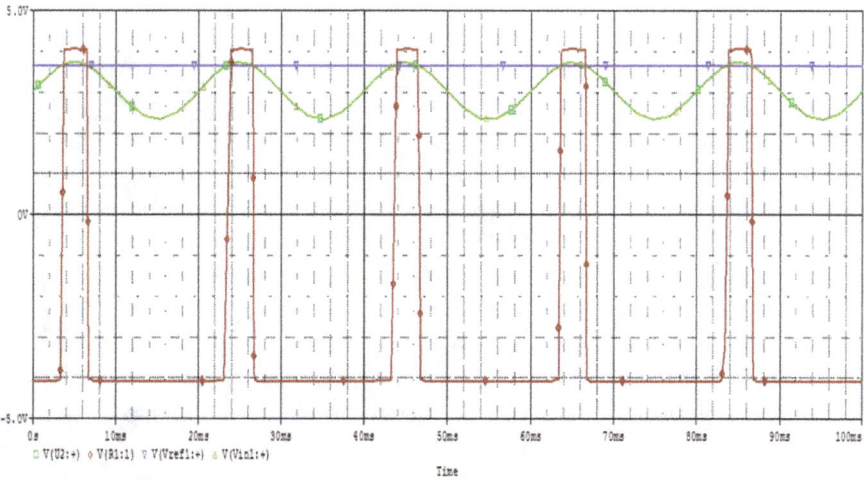

Figure 8 Simulation result of the arc fault temperature detector.

and 3.0 V for pressure sensor. The two square waves are the output from voltage comparator which trigger as logic '1' ('HIGH') at 8.5 V if the input voltage of sine wave is higher than the reference voltage and becomes logic '0' ('LOW') at −5.0 V if the input voltage of sine wave is lower than the reference voltage.

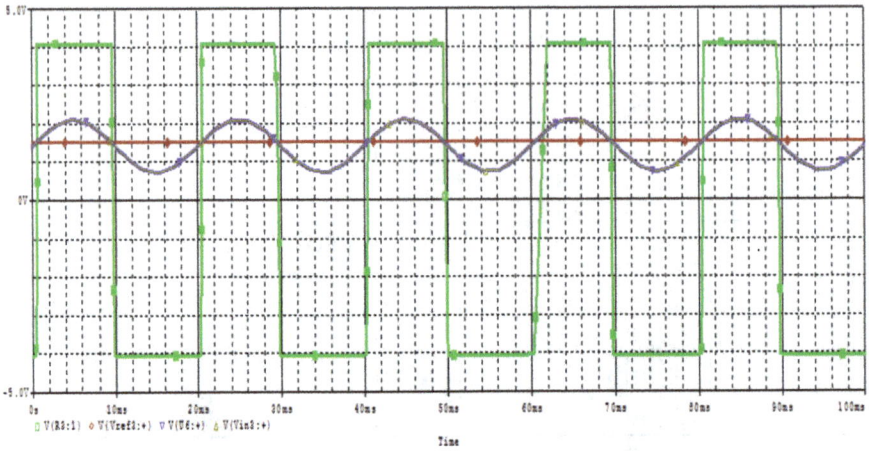

Figure 9   Simulation result of the arc fault light detector.

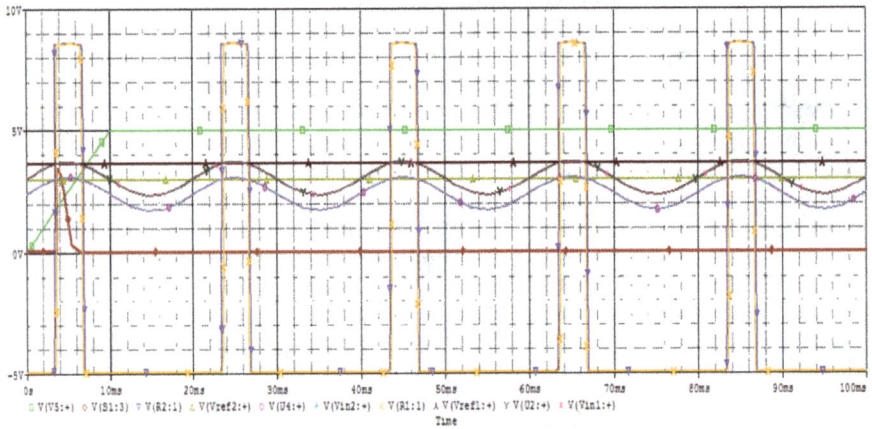

Figure 10   Simulation result of arc fault pressure and temperature detectors.

The red line at 3.5 V between the times of 0 ms to 10 ms is the output voltage of AND gate which combine the output voltage of pressure and temperature detector. After 10 ms the red line becomes 0 V as it trip the signal at the voltage controlled switch. The voltage controlled switch is act like a relay and will close when it receive a 'HIGH' output from the AND gate. Equivalently, it will trigger the voltage pulse at 5.0 V (represent in green color) which then turn on the LED.

Unfortunately, there is some limitation on the PSpice software due to the analog node limit is exceeded when we run the overall circuit with these three detectors. However, from the simulation results obtained from the individual detector, it is expected that the controlled voltage switch will receive a 'HIGH' output from the AND gate when the output from these three detectors output is sent to the 7410 AND gate. Then the voltage controlled switch will act like a relay and close to complete the circuit and allows the voltage pulse triggers at 5.0 V. Then, the LED will turn on.

6 Conclusion

Arcing fault in low voltage switchboard is a serious issue as the effects of the arcing faults are devastating. In this paper, the proposed arc fault detection circuit will activate the relay and send a trip signal to the circuit breaker if and only if all three signals, i.e. pressure, thermal and light are detected. With these three arc fault signals prior the occurrence of arc, the proposed design is able to ensure that no fault tripping signal is sent to the circuit breaker and therefore no unnecessary power shut down.

An early detection of arc fault in low voltage switchboard enable the isolation of the power supply to the consumer side just before the occurrence of arc fault and thereby reduce the danger to personal injury and building. In addition, it improves the system reliability without power interruption which is particular essential to hospitals and certain industries with sensitive loads. The circuit help to eliminate the possibility of an arc occurring and hence prevents against the effects of arc occurrence. The proposed circuit can be modeled to meet specifications of industry with different supply requirements. It is easy to design and highly reliable.

References

[1] Gammon, T., and Mattews, J. (1999) "The historical evolution of arcing-fault models for low voltage systems," in *Proceedings of the IEEE Industrial & Commercial Power Systems Technical Conference*, Detroit, MI.

[2] Kuan, L. C. (2013). Arc fault pressure detector in low voltage switchboard. *Int. J. Sci. Res. Pub.* 3.

[3] Wilson, R. A. (ABB Inc.), Harju, R. (ABB Oy, Finland), Keisala, J. (ABB Oy, Finland), and Ganesan, S. (ABB Ltd., India), "Tripping with the Speed of Light: Arc Flash Protection".

[4] Knapek, W., OMICRON Electronics Corp. Zeller, M., and Schweitzer Engineering Laboratories, Inc., (2011). *Verify Performance and Safety of Arc-Flash Detection Systems*. Pullman, WA: Schweitzer Engineering Laboratories, Inc.

[5] Land, H. B. (2008). The behavior of arcing faults in low voltage switchboards. *IEEE Trans. Ind. Appl.* 44, 437–444.

[6] Malmedal, K., and Sen, P. K. (2000). "Arcing fault current and the criteria for setting ground fault relays in solidly-grounded low voltage systems," *Proceedings of the Industrial and Commercial Power Systems Technical Conference*, Detroit, MI.

[7] Gammon, T., and Matthews, J. (2000). "Arcing fault models for low voltage power systems," in *Proceedings of the Industrial and Commercial Power Systems Technical Conference*, Detroit, MI.

[8] Kuan, L. C. (2015). "Arc fault temperature detector in low voltage switchboard," in *Proceedings of the International Conference of Information, System and Convergence Applications*, Kuala Lumpur.

[9] Baliga, B. R., and Pfender, E. (1975). *Fire Safety Related Testing of Electric Cable Insulation Materials*. Minneapolis, MN: University of Minnesota.

[10] Land, H. B., Eddins, C. L., and Klimek, J. M. (2004). *Evolution of Arc Fault Protection Technology*. Washington, DC: John Hopkins APL Technical Digest.

[11] Seedorff, J. (2015). *Arc Flash Protection: Key Considerations for Selecting An Arc Flash Relay*. Chicago, IL: Littelfuse, Inc.

[12] "Room Illumination level: General Building Areas", IES Standards, MS 1525.

Biographies

K. L. Choo received the Bachelor of Engineering from University Tenaga Nasional, Malaysia and M.Sc. degree from Imperial College, London, in 2001 and 2002 respectively. She is a Professional Engineer of Board of Engineers Malaysia (BEM) and Corporate Member of The Institution of Engineers Malaysia (IEM). Currently, she works as a lecturer in Infrastructure University Kuala Lumpur, Malaysia. Her research interests include control systems, power quality and electrical power system.

A. A. Sa'adoni received the Bachelor of Electronics Engineering (Hons) from Infrastructure University Kuala Lumpur, Malaysia in 2016. Currently, he works as a project engineer in SRS Power Engineering Sdn. Bhd. (an established low voltage switchboard company). He has major experience and knowledge in electrical power system (infrastructure), telecommunication (fiber optics) and power electronics (arc fault).

Simultaneous Integrated Model with Multiobjective for Continuous Berth Allocation and Quay Crane Scheduling Problem

Nurhidayu Idris[1] and Zaitul Marlizawati Zainuddin[2]

[1]*Department of Mathematical Sciences, Faculty of Science,*
Universiti Teknologi Malaysia, Johor, Malaysia
[2]*Department of Mathematical Sciences, Faculty of Science*
and Universiti Teknologi Malaysia – Centre of Industrial
and Applied Mathematics, Universiti Teknologi Malaysia,
Johor, Malaysia
E-mail: nurhidayu5@live.utm.my; zmarlizawati@utm.my

Received 23 October 2016; Accepted 24 November 2016;
Publication 19 December 2016

Abstract

This paper presents the simultaneous integration model of berth allocation and quay crane scheduling. Berths and quay cranes are both critical resources in port container terminals. The mathematical model uses a mixed integer linear programming with multiple objectives generated by considering various practical constraints. Small data instances have been taken to validate the integrated model. A numerical experiment was conducted by using LINGO programming software to evaluate the performance and to obtain the exact solution of the suggested model.

Keywords: Berth allocation, Quay crane scheduling, Multi objectives, Integration.

Journal of Industrial Engineering and Management Science, Vol. 1, 89–100.
doi: 10.13052/jiems2446-1822.2016.006

1 Introduction

In today's global economy, shipping, port and logistics activities are well-known and indispensable. They have an important role to play in facilitating the global trade nowadays. As we can observe, most of the businesses around the world involve shipping from one destination to other destination. Since the water transportation becomes an essential need for human daily activities especially in business field, it has caused a great demand for port container terminals. A great demand for seaport container terminals recently in business, there exist the systematic operation competition between ports. Regarding to the fierce competition among container terminals, high performance with an efficient processing in a container terminal is vital as mentioned in [1]. Previously, Daganzo [2] and Park and Kim [3] were both first presented the integrated approaches. Both papers highlighted applying monolithic model for crane assignment and crane scheduling as in [2] and berth allocation with crane assignment as in [3]. Generally, the handling time of the ship is affected by the crane schedule. The integration concept of Berth Allocation Problem (BAP) and Quay Crane Scheduling Problem (QCSP) was proposed by Park and Kim [3] with continuous berth model. The problem is formulated using the mixed integer programming (MIP) model whereby both time and space were discretized and a two phased procedure was designed to solve the problem. In the first phase, berth allocation and rough quay crane allocation was determined while in the second phase by using the solution found from the first phase, a detailed individual crane scheduling was generated. Next, this study has been extended as in [4] by restricting the moving range of each crane individually at the quay. The differences exist between these two papers are they allocate the two resources which are berths and quay cranes simultaneously, not sequentially like in [3].

In addition, the simultaneous concept has been studied as in [5] on the integration of berth allocation and quay crane scheduling problem. This problem was designed simultaneously between scheduling of cranes for all vessels and assigning of vessels to the berths. Continuous layout of berth was chosen in order to be more practical and less time consumed by vessels at port. Aykagan [5] proposed to solve the integration problem by applying a Tabu search heuristic with the objectives to minimize the serving time and weighted tardiness of vessels. Besides that, the same simultaneous concept was proposed in [6] for the integration model of BAP and QCSP. The integrated optimization model on berth allocation and quay crane scheduling was built with the objective to reduce the operating costs of quay crane and vessels.

Wang, Cao, Wang and Li [6] as well developed a new genetic and ant colony algorithm whereby the partial allocation plan is solved using genetic algorithm, then adjusted against berth through ant colony algorithm to find an optimal solution. Compared to Aykagan [5], Wang, Cao, Wang and Li [6] focused on a discrete berth layout rather than continuous berth. However, both papers focused on a dynamic concept for the arrival of ships. The integration of BAP and QCSP also has been studied as in [1] on a simultaneous process. The study is solved in a simultaneous way with uncertainties on arrival time of vessel and container handling time. The spatial and temporal constraint applied in [1] is similar to [6] whereby discrete layout of berths with dynamic arrival of vessels. A simulation based on Genetic Algorithm (GA) search procedure is built in order to generate robust berth and QC schedule proactively.

2 Problem Description

Mainly, the focus in this research is to develop a simultaneous Integration Continuous Berth Allocation Problem and Quay Crane Scheduling Problem (IBAPCQCSP) while considering multiple objectives. This paper involves two processes namely, berth allocation and quay crane scheduling which are vital operations in container terminal. Thus, the flow of the operations has been presented in the mathematical model throughout this paper. The handling time of vessel depends on the resources utilization efficiency which may result an early departure of vessel. This research focuses on continuous berth layout whereby the vessels can be served wherever empty spaces are available along the quay. Next, to formulate the multi objectives model of simultaneous IBAPCQCSP, below assumptions need to be followed:

1. The length of each vessel represents through number of holds of 2, 3 or 4 holds whereby only one crane can work on a hold at one time.
2. Numbers of vessels can berth along the quay and being served immediately.
3. During the planning horizon, vessels can arrive at port and cannot be served before its arrival.
4. Once all holds on a vessel is processed by quay cranes, a vessel is completed.
5. Once the cranes start to work on a hold, the work has to be continued until the hold completely being served.
6. Quay cranes cannot cross each other while on the same tracks.

7. After the process of loading and unloading container is completed on all holds, only then a vessel can leave the port.
8. Handling time of vessel depends on the number of quay cranes assigned to the vessel.
9. Quay cranes can move from hold to hold either in the same or different vessel without crossing one another.

3 Mathematical Formulation

In this section, a multiple objectives mathematical model is formulated with the simultaneous integrated berth allocation and quay crane scheduling. The study focuses on the dynamic arrival of vessels where a set V of vessels with known arrival times, whereby $n = |V|$. For each vessel $i \in V$, the parameter and formulation applied in this study is defines as below:

L: set of berths along the quay

Q: a set of non-identical quay cranes working on a single set of rails.

n_q: travel time of quay crane q,

T: time period of vessels

V: a set of vessels

M: a large scalar number

h_i: number of holds of vessel i,

pr_i^k: processing time of hold k of vessel i,

pr_i^{max}: maximum hold processing time for vessel i

pr_i^{max}: $\max_i \{pr_i^k\}$

f_i: lateness penalty time of vessel i,

d_i: due time of vessel i (where d_i, $a_i + pr_i^*$)

a_i: arrival time of vessel i,

bp_i: berthing position of vessel i,

bt_i: berthing time of vessel i,

e_i: the earliest time that vessel i can depart.

$$
x_{ij} = \begin{cases} 1 & \text{if vessel } i \text{ berth after vessel } j \text{ departs,} \\ 0 & \text{otherwise;} \end{cases}
$$

$$
y_{ij} = \begin{cases} 1 & \text{if vessel } i \text{ berth completely above vessel } j \text{ on the} \\ & \text{time-space diagram,} \\ 0 & \text{otherwise;} \end{cases}
$$

$$
r_{iqk}^t = \begin{cases} 1 & \text{if at least one quay crane is assigned on hold } k \text{ of vessel} \\ & i \text{ at time } t, \\ 0 & \text{otherwise;} \end{cases}
$$

$$
w_i^t = \begin{cases} 1 & \text{if at least one quay crane is assigned to vessel } i \text{ at time } t, \\ 0 & \text{otherwise;} \end{cases}
$$

The simultaneous concept of IBAPCQCSP model with multiple objectives and dynamic arrival of vessels is formulated as follow:

$$
Min \sum_{i=1}^{n} (e_i - a_i) + \sum_{i=1}^{n} f_i (e_i - d_i) + \sum_{i=1}^{n} \sum_{t \in T} \sum_{q \in Q} \sum_{k=1}^{h_i} n_q r_{iqk}^t \qquad (1)
$$

subject to

$$
x_{ij} + x_{ji} + y_{ij} + y_{ji} \geq 1 \quad \forall i, j \in V \text{ and } i < j \qquad (2)
$$

$$
x_{ij} + x_{ji} \leq 1 \quad \forall i, j \in V \text{ and } i < j \qquad (3)
$$

$$
y_{ij} + y_{ji} \leq 1 \quad \forall i, j \in V \text{ and } i < j \qquad (4)
$$

$$
bt_j \geq e_i + (x_{ij} - 1)M \quad \forall i, j \in V \text{ and } i < j \qquad (5)
$$

$$
bp_j \geq bp_i + h_i + (y_{ij} - 1)M \quad \forall i, j \in V \text{ and } i < j \qquad (6)
$$

$$
bt_i \geq a_i \quad \forall i \in V \qquad (7)
$$

$$
bt_i \geq tr_{iqk}^t + (1 - r_{iqk}^t)T \quad \forall i \in V, \forall k \in \{1, \ldots, h_i\}, \qquad (8)
$$
$$
\forall t \in T
$$

$$
e_i \geq tr_{iqk}^t + pr_i^k \quad \forall i \in V, \forall k \in \{1, \ldots, h_i\}, \qquad (9)
$$
$$
\forall t \in T, \forall q \in Q
$$

$$
e_i \geq bt_i + pr_i^{max} \quad \forall i \in V \qquad (10)
$$

$$
w_i^{min} \leq \sum_{i \in V} r_{iqk}^t \leq w_i^{max} \quad \forall q \in Q, \forall k \in \{1, \ldots, h_i\}, \qquad (11)
$$
$$
\forall t \in T
$$

$$
\sum_{i \in V} \sum_{q \in Q} \sum_{k=1}^{h_i} r_{iqk}^t \leq Q \quad \forall t \in T \qquad (12)
$$

$$\sum_{q \in Q} r_{iqk}^t = w_i^t \quad \forall i \in V, \ \forall k \in \{1, \ldots, h_i\}, \qquad (13)$$

$$\forall t \in T$$

$$\sum_{t \in T} w_i^t = e_i - bt_i \quad \forall i \in V \qquad (14)$$

$$1 \le bp_i \le L - h_i + 1 \quad \forall i \in V \qquad (15)$$

$$x_{ij} \in \{0,1\}, \ y_{ij} \in \{0,1\} \quad \forall i, j \in V \text{ and } i \ne j \qquad (16)$$

$$r_{iqk}^t \in \{0,1\}, \ w_i^t \in \{0,1\} \quad \forall i \in V, \ \forall k \in \{1, \ldots, h_i\}, \qquad (17)$$

$$\forall t \in T, \ \forall q \in Q$$

Referring to mathematical formulation above, the first objective that needs to be achieved is to minimize the duration of vessels at port and second objective is to reduce the waiting time of vessels to be served. The last objective is to lessen the travelling time of quay cranes. Constraints (2) through (4) are applied to prevent the overlapping rectangle of vessels. Constraints (5) and (6) ensure that the chosen berthing times and berthing positions are consistent with the definitions of x_{ij} and y_{ij} with M as a large positive number. Constraint (7) forces the berthing of vessels must be after their arrivals and constraint (8) forces the work on hold of vessels to begin immediately after berthing. Constraint (9) ensures that vessel can leave the port after all holds finished processing by quay cranes. Constraint (10) is to provide the lower bound on the completion time of vessel, e_i given berthing time, bt_i. Constraint (11) ensures that only one to four quay cranes can serve a vessel at the same time. Constraint (12) ensures that the working quay cranes at any time period cannot exceeded Q, and constraint (13) ensures the consistency between w_i^t and r_{iqk}^t whereby the number of quay cranes operated on vessel i at time t are equal to the summation of quay cranes assigned to the holds of vessel i at time t. Constraint (14) sets the number of quay cranes work for the vessels must be equal to the duration time of vessels being served. Constraint (15) is to make sure that the vessels fit on berths along the quay.

3.1 Simple Numerical Example

The model formulation presented previously was validated by using commercial programming software, LINGO 15.0 for small problem which can be seen in Table 1 with the length of berth, $L = 7$ and quay cranes, $Q = 4$.

Table 1 A small instance for simultaneous IBAPCQCSP

i	1	2	3	4	5
a_i	2	1	3	1	4
d_i	8	4	9	5	11
f_i	3	4	3	3	4
h_i	2	3	3	4	4
pr_i^1	3	2	2	3	3
pr_i^2	4	2	4	1	2
pr_i^3	–	2	1	0	2
pr_i^4	–	–	–	1	4

Table 2 A travel time of quay cranes (hours)

q	1	2	3	4
n_q	0.10	0.20	0.15	0.25

In this study, the length of vessel is represented through the different number of holds for each vessel which are 2, 3 or 4 holds. The vessels can moor at the berth along the quay and obtain service by quay cranes simultaneously with only one crane can assign and works on a hold. The processing time of each hold has been assigned for every vessel to be served by quay cranes. The quay crane is allowed to shift form one hold of a vessel to another only if a quay crane finished processing initially assigned hold. Each vessel is considered completed once the quay cranes finished processing all holds in a vessel. After all the vessels are finished servicing, the exact solution is obtained and presented through a time-space diagram or Gantt chart which is the best way to represent a vessel who is queuing to be served and in the middle of servicing.

3.2 Main Results

From the simple numerical data assigned, the exact solutions by using three different rules are compared. Table 3 and Figure 1 shows the exact solution of using First Come First Serve (FCFS) rule while Table 4 and Figure 2 shows the result obtained by applying Large Vessel First (LVF) rule. For the last result, Shortest Processing Time First (SPTF) rule is applied and the exact solution can be seen in Table 5 and Figure 3 as followed. The moving path of the quay cranes also will be presented in the time space diagram.

Based on the three exact solutions diagrams above, the total objective function value obtained by applying FCFS rule was 40.4 hours with the idle cranes of 8 while LVF rule with total optimal objective function value of

Table 3   The exact solution for simultaneous IBAPCQCSP (FCFS rule)

i	1	2	3	4	5
bp_i	1	1	5	4	1
bt_i	3	1	6	1	7
e_i	7	3	11	6	12
r_{iq}^1	3	1	6	1	7
r_{iq}^2	3	1	6	4	7
r_{iq}^3	–	1	10	–	9
r_{iq}^4	–	–	–	5	8

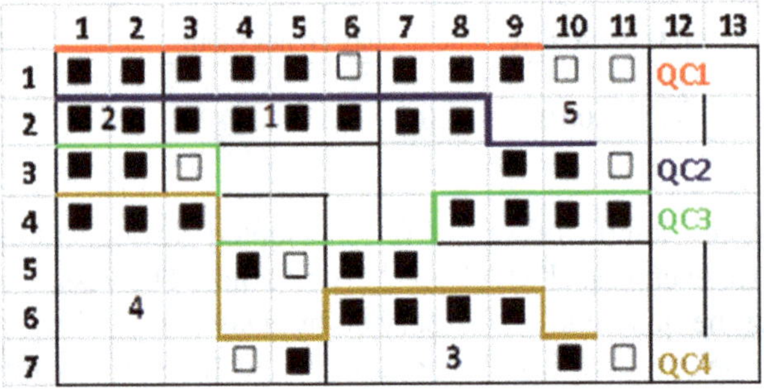

Figure 1   A time space diagram for the exact solution of simultaneous IBAPCQCSP (FCFS rule).

Table 4   The exact solution for simultaneous IBAPCQCSP (LVF rule)

i	1	2	3	4	5
bp_i	1	5	5	1	1
bt_i	8	2	4	1	4
e_i	12	4	10	4	8
r_{iq}^1	8	2	8	1	4
r_{iq}^2	8	2	4	1	4
r_{iq}^3	–	2	8	–	6
r_{iq}^4	–	–	–	1	4

33.2 hours and 8 idle cranes. For SPTF rule, the total objective function value obtained was 28.35 hours with 8 idle cranes too. It shows that the performance of the integrated operation regarding BAP and QCSP by using the SPTF rule is smaller compared to FCFS and LVF rules. Therefore, the SPTF rule is more applicable and effective for the model to be used in further research. From the results, SPTF rule gives the lowest results since the berth was fully occupied with vessels from time to time and there was lessen delay completion time of vessels.

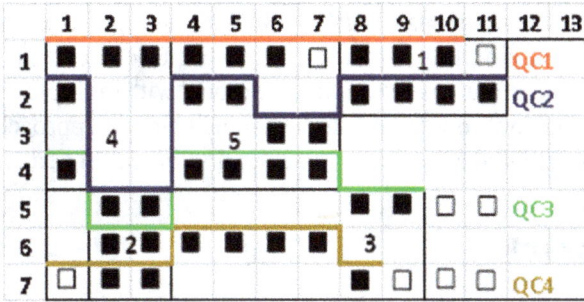

Figure 2 A time space diagram for the exact solution of simultaneous IBAPCQCSP (LVF rule).

Table 5 The exact solution for simultaneous IBAPCQCSP (SPTF rule)

i	1	2	3	4	5
bp_i	5	1	1	4	1
bt_i	5	1	3	1	7
e_i	9	3	7	5	12
r_{iq}^1	5	1	3	1	7
r_{iq}^2	5	1	3	4	7
r_{iq}^3	–	1	3	–	9
r_{iq}^4	–	–	–	4	8

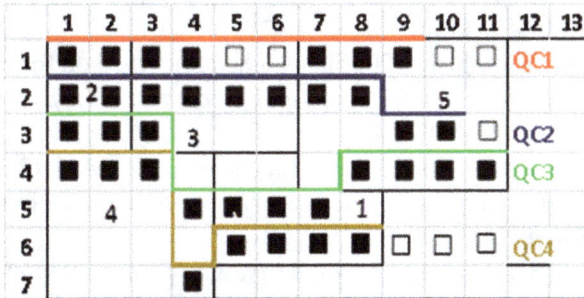

Figure 3 A time space diagram for the exact solution of simultaneous IBAPCQCSP (SPTF rule).

4 Conclusion

The highlight of this paper is the integration on operating both resources which are berths and quay cranes at the port terminal simultaneously. There exist two processes namely, BAPC and QCSP which increase the operations efficiency at container terminal. This paper proposed a mathematical formulation for the simultaneous IBAPCQCSP with multiple objectives. This study also performed simple numerical experiment to evaluate the performance and validate the simultaneous integrated model with multiple objectives added using LINGO 15.0. For future studies, metaheuristic approach will be considered for the simultaneous integrated model with real life data.

Acknowledgment

The authors would like to thank the Malaysian Ministry of Higher Education for their financial funding through Fundamental Research Grant Scheme (FRGS) vote number R.J130000.7809.4F470. This support is gratefully acknowledged. The author would also like to thank UTM Centre for Industrial and Applied Mathematics and Department of Mathematical Sciences, Faculty of Science, UTM for the support in conducting this research work.

References

[1] Han, X. L., Lu, Z. Q., and Xi, L. F. (2010). A proactive approach for simultaneous berth and quay crane scheduling problem with stochastic arrival and handling time. *Eur. J. Operat. Res.* 207.3, 1327–1340.
[2] Daganzo, C. F. (1989). The crane scheduling problem. Transport. Res B Methodol. 23.3, 159–175.
[3] Park, Y. M., and Kim, K. H. (2005). "A scheduling method for berth and quay cranes." in *Container Terminals and Automated Transport Systems.* Berlin: Springer, 159–181.
[4] Zhang, C., Zheng, L., Zhang, Z., Shi, L., and Amstrong A. J. (2010). The allocation of berths and quay cranes by using a sub-gradient optimization technique. *Comput Ind. Eng.* 58.1, 40–50.
[5] Aykagan, A. (2008). *Berth and quay crane scheduling: problems, models and solution methods.* ProQuest, Georgia Institute of Technology, Georgia.
[6] Wang, R. D., Cao, J. X., Wang, Y., and Li, X. X. (2014). An integration optimization for berth allocation and quay crane scheduling method based on the genetic and ant colony algorithm. *Appl. Mech. Mater.* 505–506, 940–944.

Biographies

N. Idris is a master student at the Universiti Teknologi Malaysia at Skudai, Johor, Malaysia since 2014. She attended the Universiti Teknologi Malaysia, Johor where she received her B.Sc. in Industrial Mathematics in 2014. Nurhidayu is currently completing a master' degree in Applied Mathematics specialized in Operational Research field at the Universiti Teknologi Malaysia at Skudai, Johor, Malaysia. Her master work discusses on minimizing the operation at port which involving Berth Allocation and Quay Crane Scheduling Problem.

Z. M. Zainuddin is a Senior Lecturer at Universiti Teknologi Malaysia, Malaysia. She received his B.Sc. degree in Mathematics from University of Southampton, United Kingdom in 1996, M.Sc. degree in Operational Research and Applied Statistics, from Salford University, United Kingdom, in 1999, and his Ph.D. degree in Mathematics and Statistics from University of Birmingham, United Kingdom, in 2004. Her research interest includes modeling and solving real life problem using Operational Research, Optimization and Heuristic techniques. She is a member of the OR Society, Management Science/OR Society of Malaysia (MSORSM) and Malaysian Mathematical Sciences Society (PERSAMA).

Monitoring the Level of Light Pollution and Its Impact on Astronomical Bodies Naked-Eye Visibility Range in Selected Areas in Malaysia Using Sky Quality Meter

M. S. Faid[1,*], N. N. M. Shariff[1], Z. S. Hamidi[2], S. N. U. Sabri[2], N. H. Zainol[2], N. H. Husien[2] and M. O. Ali[2]

[1]Academy of Contemporary Islamic Studies, Universiti Teknologi MARA, Shah Alam, Selangor, Malaysia
[2]Faculty of Applied Sciences, Universiti Teknologi MARA, Shah Alam, Selangor, Malaysia
E-mail: msyazwanfaid@gmail.com; nur.nafhatun.ms@gmail.com; zetysh@salam.uitm.edu.my; sitinurumairah@yahoo.com; hidayahnur153@yahoo.com; hazwani_husien21@yahoo.com; marhanaomarali@gmail.com
*Corresponding Author

Received 20 October 2016; Accepted 21 November 2016; Publication 8 December 2016

Abstract

Light pollution is an anthropogenic by-product of modern civilization and heavy economical activity, sourced from artificial light. In addition of its detrimental impact on human and ecology, light pollution brightens the night sky, limiting the range of visible astronomical bodies detected by naked-eye. Since it is becoming a global concern for astronomers, the level of light pollution needs to be monitored to study its mark on the astronomical data. Using Sky Quality Meter in the period of 5 months, we investigated the links between city population and its vicinity from the city center towards the profile of the night sky and the limiting magnitude of the naked eye. We eliminate the data factored by clouds and moon brightness on account of it has an adverse effect on sky brightness that could disrupt research on light pollution.

Journal of Industrial Engineering and Management Science, Vol. 1, 101–118.
doi: 10.13052/jiems2446-1822.2016.007

From the result, we can see population and location distance from the city as major variables of light pollution, as Kuala Lumpur, a city center sky is 5 times brighter than Teluk Kemang, a suburban sky. Some recommendation in reducing the effect of light pollution will be also discussed.

Keywords: Anthropogenic pollution, Light pollution, Population, Night sky brightness, Limiting magnitude.

1 Introduction

One of the massive downfalls of the modern civilization is the alteration of the natural environment, which include ambient light alteration in the night sky. The man-made light, or artificial light produces a sky glow, that scattered vertically, then returned back into our sky by the atmosphere making the sky brighter and disrupts the suppose night sky ambient. This phenomenon classified as one the human pollution towards the environment and it is called light pollution [1].

Increase in human population density in a certain location parallels with growth in economic and social infrastructure. However, this unstoppable economic growth and population development crops a massive light output through their street light and giant building spotlight that is directly proportional to the light pollution [2]. This trends keeps deteriorating, without people realizing it could damage the environment and the health of living things [3]. The truth is people find other issue more important although they concerned about the environment [4].

Naturally, human need a steady cycle of day and night hours throughout their life. The exposure of the artificial light imitates the sun brightness that disrupt the efficacy of melatonin in phase shifting circadian rhythms [5]. This disruption can cause widespread interruption of multiple body systems, resulting in serious medical consequences for individuals, such as bad job performance, increase in weight and even as bad as cancer [6].

Besides human, the light pollution can also affect animal. Throughout the synodic month, the night sky brightness is shaped by rhythmical shift of celestial object, namely moon. The night sky brightness varies from highest when the full moon and lowest at the new moon. Animal, bug, and all other living things acclimatize their life activity and cycle according to the synodic month cycle. Unfortunately, when monthly periods of sky brightness is interfered by artificial light, it hampers their harmonious life cycle [1]. The nightly routine of insects [7], birds [8] and turtles [9] is threatened as

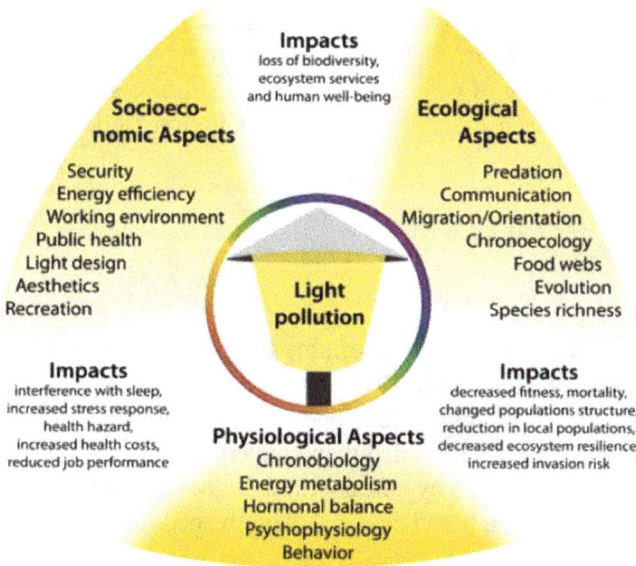

Figure 1   The effect of light pollution on living organisms [10].

it distorts their nocturnal nature [10], encompassing from foraging, mating, hatching and night navigation [11].

In terms of astronomy, the ground-observable astronomical bodies in the sky are determined by the contrast of these bodies with the night sky brightness [12]. The human eye has a certain contrast threshold [13] in detecting a dim celestial object in the sky [14]. The detection of celestial object on a certain night sky is quantified as Bortle scale [15], which believed to be application of knoll experiment. A highly polluted sky with low magnitude of sky brightness will lower the amount of celestial object visible to both naked and optical aid observation [16]. This make light pollution is the major concerns of astronomers considering its effect on ground based observation especially on optical and radio observation [17]. Therefore, it is an utmost importance to monitor the light pollution level to stress out its mark on astronomical data. According to the Garstang [18], the night sky brightness varies from one location to another location, depends on the element of population, economic activity and location distance from the nucleus of the population. A lowly populated area has a high viewing chance of multiple astronomical bodies compared to the heavily dense area [19]. Concerning about this problem, we examined the character of night sky brightness in selected areas in Malaysia. The purpose of the examination is to study the magnitude of sky brightness from different location profiles and to determine the number of celestial objects in the spectrum of naked eye visibility in respect to the locations. Since the zenith sky brightness signify the level of light pollution at that particular location [20], we can see the pattern of light pollution from two different location profile.

2 Methodology

2.1 Instrument

In this study, we will use SQM to determine the sky brightness of the night sky. The Sky Quality Meter (SQM) is a pocket device, to evaluate the sky brightness in the unit of mag/sec^2 developed by Unihedron. SQM use a light intensity to frequency detector, which then convert the output in digital LED reading. The sensor is covered by a near-infrared blocking filter to make sure the brightness reading is on the visible wavelength spectrum.

$$fo = Fd + (Re)(Ee) \qquad (1)$$

Where

Fo is the output frequency,
Fd is the output frequency for dark condition
Re is the device responsivity for a given wavelength of light given in kHz/(μW/cm^2)
Ee is the incident irradiance in μW/cm^2

The output frequency is then converted into magnitude in the mean of formula

$$m = -5/2(\log(fv/Jy)) + 8.9 \qquad (2)$$

Where

Fv is the spectral flux density or output frequency
Jy is the constant unit of Jansky

2.2 Site Survey

In this study, we stationed the SQM in two locations that have different site profile in terms of population density and its vicinity from the city center. These locations are classified as urban and suburban site, named Kuala Lumpur and Teluk Kemang respectively, and been pinpointed on the map by white mark. Figure 2 demonstrates the radiance map layered with population, courtesy

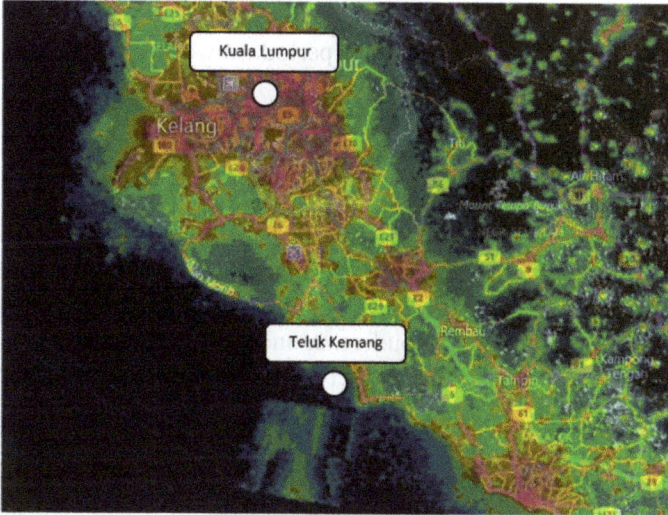

Figure 2 Irradiance mapping of Kuala Lumpur and Teluk Kemang.

of Jurij Stare[1] which is from night time radiance composite images using Visible Infrared Imaging Radiometer Suite (VIIRS) Day/Night Band (DNB) produced by the Earth Observation Group, NOAA National Geophysical Data Centre. As you can see from the map, Kuala Lumpur is located at the center of the city, with a population of 1,674, 621 people in an area of 242.7 km^2, meaning it has the density of 6900 inh/km^2. While Teluk Kemang is located just at the 4 km from the city center Port Dickson, with a population of 115 361 in an area of 575.76 km^2, indicates it has the density of 200.4 inh/km^2. To monitor the overall light pollution, the measurement of night sky brightness is taken in the period of 5 months, from August to December.

Ten nights of hourly data are taken per month in Kuala Lumpur as heavy population traffic and close proximity to city suggest an inconsistency in sky magnitude thus longer hours of monitoring is needed for calibration to yield the actual night sky brightness data. On the other hand, single night per month is enough to represent the night sky brightness in Teluk Kemang, since it has a far distance from the city nucleus and dormant human activity.

Besides site distance from the city and its population density, there a few variables that could deviate the magnitude reading from the actual sky brightness. These variables occur naturally to any place in the world, namely the amount of cloud in the sky and the brightness of the Moon. Considering

[1]www.lightpollutionmap.info

Malaysia climatology and the moon brightness [21], we expected the data to be fluctuated throughout measurement period, however, this data can be calibrated with a series of step.

2.3 The Sky Brightness Fluctuation from Cloud Cover Amount

One of the important aspects that could affect the sky brightness is the amount of cloud cover. In the case of Malaysia, the magnitude inconsistency of sky brightness is the result of the natural climatology of cloud in Malaysia which have a 65–95 percent amount of cloud cover in the sky throughout the year [22]. An overcast sky has a massive impact on the brightness of the night sky ranging from 3 to 18 times darker than clear sky depending on the location profile [23].

The high reflectivity of cloud reduces the amount of incoming solar radiation absorbed by the Earth-atmosphere system by increasing the albedo. Henderson describes the relationship between cloud and radiation of the sky by [24]:

$$Net = S(1 - a) - F \qquad (3)$$

Where the net is the net radiation at the top of the atmosphere, S is incoming solar radiation or artificial light, α is albedo of the earth-atmosphere, and F is the infrared emission of space by the system. This equation indicates that an increase of cloud amount, will increase the albedo, which in our cases the sky brightness. We implemented the Okta Scale to study the effect of the cloud amount to the brightness of the sky in Okta Scale in 8-point scale ranging from 0 (completely clear sky) to 8 (fully covered sky). A study by Kocifaj and Solano Lamphar simulate that distance plays an important part for night sky irradiance due to cloud cover [25], and since Teluk Kemang is far from the city center for its night sky to be effected by the cloud, we do not include the analysis of Teluk Kemang's Okta Scale.

2.4 Effect of Moon Phase on Sky Brightness

After the effect of the cloud, the Moon is the highest natural contributor of the sky brightness. The Moon's scattered radiation depends on its altitude and phases, where we calculate the elliptical coordinate from Jean Meeus algorithms [26] and the moon phase is calculated from Smith [27]. Jason formulated that the moon scattered radiation can influences the night sky brightness up to 7 magnitude difference [28] which when full moon at its maximum brightness. A moon phases more than 0.5 in has long hour regime in the night sky while the full moon (phases > 0.9) is above the horizon all night. In this study, we will examine the magnitude reading of the night sky that has the lowest yield of sky brightness from average magnitude reading and eliminate the data that show a clear influence of moon brightness, when moon phase > 0.5.

2.5 Naked Eye Limiting Magnitude

By eliminating the data affected by cloud cover and moon phases, we then can obtain to actual yield of night sky brightness on that particular location. The difference pattern of night sky brightness at both location is then determined and the naked-eye range of visible astronomical bodies. The determination of the visible celestial object by the naked eye is expressed in the log of the night sky brightness magnitude (m) with formula [29] derived from Knoll, where NELM is the naked eye limiting magnitude.

$$\text{NELM} = 7.93 - 5 \times \log(10^{4.316-(m/5)}) + 1 \tag{4}$$

From this formula we can predict the visible object in the sky theoretically in respect of the distinction of both site profiles. The prediction is based on the Bortle Scale theory, relevant to the location characteristic.

3 Results

3.1 Light Pollution Data

The trends of the sky brightness at both stations throughout August until December are shown in Table 1. Cloud and moonlight are the major factors for the fluctuation trend of the graph. Malaysia was struck by a massive haze by November and September but the effect of haze can be accounted as same as the effect of cloud cover. Both suburban and urban sites indicate a stable difference between both of them, with urban sky brightness is 2 to 4 magnitudes brighter relative to the suburban site, under the same month and the same meteorological conditions. At a glance, we can agree that the population and their artificial light play a major explanation of the difference in magnitude, evidently shows that it is the cause of light pollution.

Since August 2015 until December 2015, a total of 55 raw data from Teluk Kemang and Kuala Lumpur. Of the Sky Brightness data collected, 50 data came from urban station Kuala Lumpur, while the rest is from suburban station Teluk Kemang. Table 2 presents the combined statistic. The total data average

Table 1   Sky brightness throughout 5 months

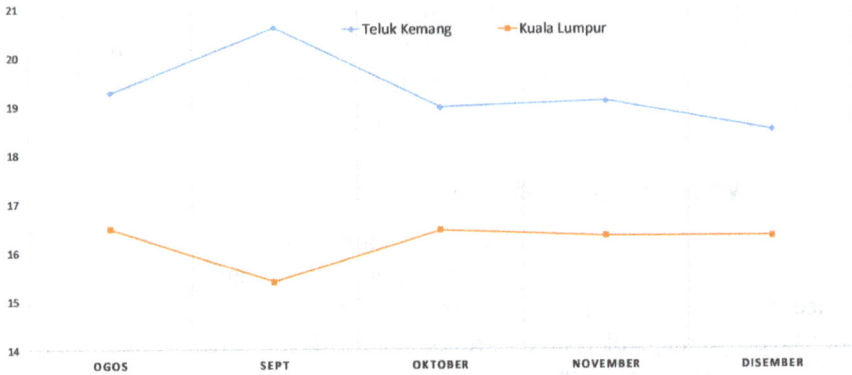

Table 2   Sky brightness data

Location (Name)	Raw Data	Average	Standard Deviation
Urban (KL)	50	16.19	0.45
Suburban (TK)	5	19.30	0.78
Total	55	16.20	0.91

is 16.2. As mentioned before, the sky brightness at the urban location show a lower average magnitude, therefore the sky brightness is higher or brighter than suburban location. The mean difference of magnitude from urban location Kuala Lumpur, with average 16.19 and suburban location Teluk Kemang, with average 19.30 is 3.11.

This implies the urban location Kuala Lumpur is 17 times brighter than rural location Teluk Kemang.[2] The combined histogram of Night Sky Brightness recorded in Figure 3. Since data from suburban location Teluk Kemang is a lot less than data from urban location Kuala Lumpur, it is being scaled so that it is comparable relatively. The largest difference of sky brightness in magnitude from urban location and suburban is 6 mag/sec^2 or 251 brighter in observed luminosity. The highest frequency at urban location Kuala Lumpur is around 16 mag/sec^2, while at suburban location Teluk Kemang is 19 mag/sec^2. Both of these values are affected by the natural climatology of Malaysia that has a high amount of cloud cover in the sky throughout the year.

---

[2]Magnitude in brightness operates in log by the base of 2.512. Therefore, the difference in mag, x, refers to flux ration of 2.512^x.

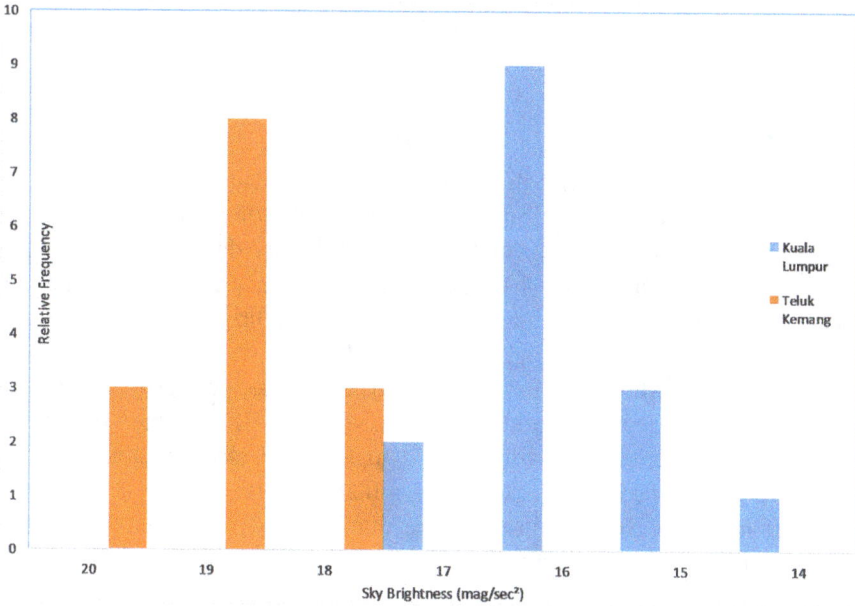

Figure 3   Relative frequency of brightness magnitude.

3.2 Cloud Cover Effect

The Okta relationship with night sky brightness is being portrayed is Table 3. At lower magnitude, which less than 17, the data is densely distributed at 6–8 Okta Scale, while at magnitude higher than 17, we can see the data sparsely dispersed at 2–5 Okta Scale. The lowest value of sky brightness is 17.59 on the Okta Scale 2 or 31 percentage of cloud cover. A full clouds cover in the sky will contribute around 1.62 difference in sky brightness or 4.4 brighter

Table 3   Average magnitude over Okta scales

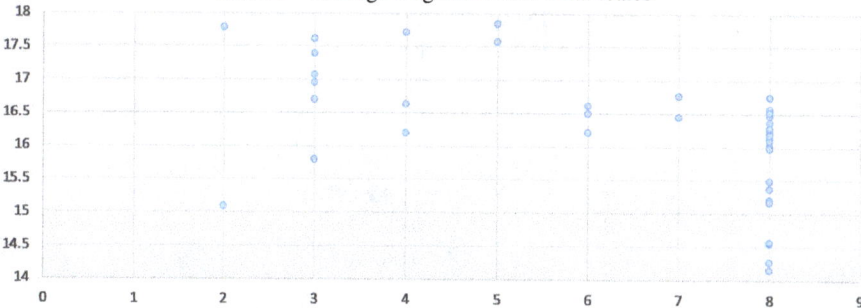

than cloudless sky. From a total of 55 data of sky brightness, a few data with magnitude less than 16 come to our attention.

3.3 Moon Phase Effect

The altitude of the moon during night depends on the phases of the moon. During the highest phases, or full moon, the moon altitude is always above the horizon throughout the night, making sky brightness data collection is not ideal. Table 4 exhibits that 4 of our data was clearly affected by the full moon, bringing a massive 2 magnitude difference in brightness from the lowest count of sky brightness, 17.59, or 6.5 times brighter. In early September until mid-November, Malaysia climatology is being stricken by terrible haze that obstructs optical depth of seeing. The aerosol particles contained in the haze scattered and absorb the artificial light making our sky brighter. This explains why data 5–10 have magnitude less than 16 even though it is not at the full moon. The effect of haze, at the maximum Air Pollution Index has contributed 3.52 differences of magnitude or 25 times brighter than the normal sky.

3.4 Naked-Eye Limiting Magnitude

By studying the natural phenomenon can that affect the sky brightness, at a normal percentage of cloud cover and neglecting the data at moon phases larger than 0.5, we can agree that 17.59 is the average brightness of the normal sky at urban location Kuala Lumpur and 19.28 is the average brightness of the normal sky is suburban location Teluk Kemang, portrayed in Table 5.

Table 4   Sky brightness & moon phases

No.	Mag.	Moon Phases
1	15.97	0.99
2	15.09	0.98
3	15.16	0.97
4	15.97	0.94
5	15.78	0.27
6	14.26	0.24
7	14.55	0.16
8	15.37	0.09
9	14.55	0.06
10	15.21	0.02

Table 5   Magnitude of brightness and NELM

Sites	Magnitude and Standard Deviation	NELM	Milky Way	Astronomical Object and Constellations
Kuala Lumpur	17.59/0.92	3.78	Not Visible at all	The Pleiades Cluster is visible, but very few other objects can be detected. The constellation is dimmer and lack of main star
Teluk Kemang	19.28/0.70	4.98	Not Visible at all	The Pleiades Cluster is the only object visible to all except for experience observer. Only the brightest constellation is discernible and they are missing star.

This imply that the night sky in Kuala Lumpur is 5 times brighter than night sky in Teluk Kemang. The brightness of the sky can affect the limiting amount of celestial object brightness visible in that particular sky, or Naked Eye Limiting Magnitude (NELM).

4 Discussion & Conclusion

After the elimination of natural contributors of night sky brightness, we can see the toll of artificial light in observing celestial objects in the sky. Kuala Lumpur sky, the site that has denser population and located in the city center is heavily polluted with magnitude value of 17.59 mag/sec^2 and 3.78 naked-eye limiting magnitude, are 5 times brighter than Teluk Kemang which has brightness of 19.28 mag/sec^2 and 4.98 naked-eye limiting magnitude.

The light pollution level in Teluk Kemang starting to worsen. Teluk Kemang has been a hot observation spot since years back and numbers of new moon observation records was achieved[3].

But with increasing human activity in its nearby city Port Dickson, sparked by Port Dickson strategic location for tourism activity and all over the world, this location will be example of many other locations that may someday

---

[3]Youngest Moon Age and Smallest Elongation. For more info, see http://www.icoproject.org/record.html

no longer viable for astronomical observation. The idea of maintaining the night sculpture in the sky may not be well accepted by the vast dimension of human society, since the impetus of humanity is driven by industrial expedience and economy, but we regard the idea of conserving mortal well-being and ecosystem stability is an idea shared by all. Attention concerning light pollution regulation and policy should be put at the highest priority since a rapid increase of artificial light every year endangering all livings things. Such simple initiative that can be implemented is by introducing the dark sky location specifically for astronomical observation and nocturnal animal activity, with a minimum distance of 7 km is good enough to conserve the night sky natural brightness. A more dramatic action is by proposing a total ban of man-made light at wavelength shorter than 540, since smaller wavelength in that spectrum has an inimical effect on human and animal health.

Acknowledgment

This work was supported by 600-RMI/RAGS 5/3 (121/2014), 600-RMI/RACE 16/6/2 (4/2014) and 600-RMI/FRGS 5/3 (135/2014) FRGS/2/2014/ST02/UITM/02/1 and Kementerian Pengajian Tinggi Malaysia. Special thanks to the Jurij Stare of www.lightpollutionmap.info and the NOAA National Geophysical Data Center for providing thematic map of night time irradiance. Also credit to Malaysia Department of Statistic for population density data and Permata Pintar Observatory UKM for atmospheric extinction observation and moon brightness data.

References

[1] Davies, T. W., Bennie, J., Inger, R., and Gaston, K. J. (2013). Artificial light alters natural regimes of night-time sky brightness. *Sci. Rep.* 3, 1722.
[2] Walker, M. F. (1977). The effects of urban lighting on the brightness of the night sky. *Publ. Astron. Soc. Pac.* 89, 405–409.
[3] Cinzano, P., and Elvidge, C. D. (2004). Night sky brightness at sites from DMSP-OLS satellite measurements. *Mon. Not. R. Astron. Soc.* 353, 1107–1116.
[4] Shariff, N. N. M., Hamidi, Z. S., Musa, A. H., Osman, M. R., and Faid, M. S. (2015). Creating awareness on light pollution' (CALP) project as a platform in advancing secondary science education," in *Proceedings of the International Conference of Education, Research and Innovation,* Seville.

[5] Burgess, H. J., Sharkey, K. M., and Eastman, C. I. (2002). Bright light, dark and melatonin can promote circadian adaptation in night shift workers. *Sleep Med. Rev.* 6, 407–420.

[6] Navara, K. J., and Nelson, R. J. (2007). The dark side of light at night: physiological, epidemiological, and ecological consequences. *J. Pineal. Res.* 43, 215–224.

[7] Frank, K. (2006). Effects of artificial night lighting on moths. *Ecol. Cons. Art. Night Light.* 2006, 305–344.

[8] Merkel, F. R., and Johansen, K. L. (2011). Light-induced bird strikes on vessels in Southwest Greenland. *Mar. Pollut. Bull.*, 62, 2330–2336.

[9] Witherington, B. E., and Bjorndal, K. A. (1991). Influences of artificial lighting on the seaward orientation of hatchling loggerhead turtles Caretta caretta. *Biol. Conserv.* 55, 139–149.

[10] Hölker, F. Wolter, C., Perkin, E. K., and Tockner, K. (2010). Light pollution as a biodiversity threat. *Trends Ecol. Evol.* 25, 681–682.

[11] Kyba, C. C. M., Ruhtz, T., Fischer, J., and Hölker, F. (2011). Lunar skylight polarization signal polluted by urban lighting. *J. Geophys. Res. Atmos.* 116, 1–7.

[12] Schaefer, B. E. (1993). Astronomy and the limits of vision. *Vistas Astron.* 36, 311–361.

[13] Tousey, R., and Hulburt, E. O. (1948). The visibility of stars in the daylight sky. *J. Opt. Soc. Am.* 38, 886–896.

[14] Crumey, A. (2014). Human contrast threshold and astronomical visibility. *Mon. Not. R. Astron. Soc.* 442, 2600–2619.

[15] Bortle, J. E. (2001). Introducing the bortle dark-sky scale. *Sky Telesc.* 60, 126–129.

[16] Falchi, F., Cinzano, P., Duriscoe, D., Kyba, C. C. M., Elvidge, C. D., Baugh, K., et al. (2016). The new world atlas of artificial night sky brightness. *Sci. Adv.* 2:6.

[17] Hamidi, Z. S., Abidin, Z. Z., Ibrahim, Z. A., and Shariff, N. N. M. (2011). "Effect of light pollution on night sky limiting magnitude and sky quality in selected areas in Malaysia," in *Proceedings of the 3rd International Symposium & Exhibition in Sustainable Energy & Environment (ISESEE)* (Rome: IEEE) 233–235.

[18] Garstang, R. H. (1991). Light Pollution Modeling. *Light Pollut. Radio Interf. Sp. Debris* 17:56.

[19] Cinzano, P., Falchi, F., and Elvidge, C. D. (2000). Naked eye star visibility and limiting magnitude mapped from DMSP-OLS satellite data. *Mon. Not. R. Astron. Soc.* 323:15.

[20] Cinzano, P., Falchi, F., and Elvidge, C. D. (2001). The first world atlas of the artificial night sky brightness. *Mon. Not. R. Astron. Soc.* 328, 689–707.

[21] Pun, C. S. J., So, C. W., and Wong, C. F. (2012). "The night sky monitoring network in hong kong," in *Proceedings of the Light Pollution: Protecting Astronomical Sites and Increasing Global Awareness through Education.*

[22] Engel-Cox, J. A., Nair, N. L., and Ford, J. L. (2012). Evaluation of solar and meteorological data relevant to solar energy technology performance in malaysia. *J. Sustain. Energy Environ.* 3, 115–124.

[23] Kyba, C. C. M., Tong, K. P., Bennie, J., Birriel, I., Birriel, J. J., Cool, A., et al. (2015). Worldwide variations in artificial skyglow. *Sci. Rep.* 5:8409.

[24] Henderson-Sellers, A., Hughes, N. A., and Wilson, M. (1987). Cloud cover archiving on a global scale: a discussion of principles. *Am. Meteorol. Soc. Bull.* 62, 1300–1307.

[25] Kocifaj, M., and Lamphar, H. A. S. (2014). Quantitative analysis of night skyglow amplification under cloudy conditions. *Mon. Not. R. Astron. Soc.* 443, 3665–3674.

[26] Meeus, J. (1991). *Astronomical Algorithms*, 2nd Edn. Virginia: Willmann-Bell.

[27] Smith, P. D. (2005). *Practical Astronomy With Your Calculator.* Cambridge: Cambridge University Press.

[28] Pun, C. S. J., So, C. W., Leung, W. Y., and Wong, C. F. (2014). Contributions of artificial lighting sources on light pollution in Hong Kong measured through a night sky brightness monitoring network. *J. Quant. Spectrosc. Radiat. Transf.* 139, 90–108.

[29] Knoll, H. A., Tousey, R., and Hulburt, E. O. (1946). Visual thresholds of steady point sources of light in fields of brightness from dark to daylight. *J. Opt. Soc. Am.* 36, 480–482.

Biographies

M. S. Faid is currently a master student at Universiti Teknologi MARA, Shah Alam Malaysia. Researching in algorithm for new moon visibility criteria. Interest in the study of light pollution and twilight sky brightness.

N. N. M. Shariff She is a senior lecturer at Academy of Contemporary Islamic Studies, Shah Alam. She obtained all her degrees from Universiti Malaya (UM) in Islamic Astronomy and Science & Technology Studies. Currently she holds several positions as a Research Management Unit (RMU) coordinator and Malaysian Research Assessment (MyRA) Liaison Officer. She is one of editorial board for Journal of Contemporary Islamic Studies (JCIS), Senior Fellow of Centre for Human Rights and Advocacy (CENTHRA) and member of Malaysian Islamic Astronomy Society. Recently she is also appointed as associate fellow at Institute of Science (IOS).

Z. S. Hamidi is currently a Senior Lecturer at Universiti Teknologi MARA, Shah Alam Malaysia. Involve in solar physics research and the project is under International Space Weather Initiative (ISWI) project under NASA project and published more than 150 publications. Also involves in others area of astronomy.

S. N. U. Sabri is currently a master student at Universiti Teknologi MARA, Shah Alam Malaysia. Involve in investigation on evaluation beta-gamma magnetic field on active region in solar radio burst.

N. H. Zainol is currently a master student at Universiti Teknologi MARA, Shah Alam Malaysia. Study the Coronal Mass Ejection Based of Moreton Waves on the Characteristic of Solar Radio Burst Type 2.

N. H. Husein is currently a master student at Universiti Teknologi MARA, Shah Alam Malaysia. Researching in solar radio burst type ii n iii related with CME and solar flare.

M. O. Ali is currently a master student at Universiti Teknologi MARA, Shah Alam Malaysia. Researching on solar radio burst type III and solar flare.

www.ingramcontent.com/pod-product-compliance
Lightning Source LLC
Chambersburg PA
CBHW061829220326
41599CB00027B/5237